Atlas of Human Embryology

Trent D. Stephens, Ph.D.
Senior Fellow, Central Laboratory for Human Embryology,
Department of Pediatrics

and

Senior Fellow, Department of Biological Structure
University of Washington
Seattle, Washington 98195

Macmillan Publishing Co., Inc.
New York
Collier Macmillan Publishers
London

To Kathy

Macmillan Publishing Co., Inc.
866 Third Avenue, New York, New York 10022

Collier Macmillan Canada, Ltd.

Library of Congress Cataloging in Publication Data

Stephens, Trent D.
Atlas of human embryology.

Includes index.
1. Embryology, Human—Atlasses. I. Title.
[DNLM: 1. Embryology—Atlases. QS617 S835a]
QM602.S74 611'.013 79-19359
ISBN 0-02-417150-6

Printing: 1 2 3 4 5 6 7 8 Year: 0 1 2 3 4 5 6

Preface

The students whom I have taught during the past few years have expressed some disappointment concerning the content of descriptive embryology laboratories. The problem primarily consists of two major points: (1) little or no human material is included, and (2) few if any embryos are examined beyond the 10-mm stage. The standard explanation is that the 10-mm pig and 10-mm human are remarkably similar in structure. While this statement is somewhat justified, no embryos are examined beyond 10-mm because there are considerable differences from that stage forward. I believe that the students' complaints are legitimate since the majority of those who take embryology are planning or involved in a career in the medical sciences.

I have found, in my experience, that the *Atlas of Descriptive Embryology* by Willis W. Mathews is the best student embryology atlas available. This *Atlas of Human Embryology* has been developed not to replace such an atlas, as the understanding of early developmental stages of other animals is important to our complete knowledge of development, but to expand and complement it by the addition of human embryos and at later stages.

The 11.5-mm human embryo was selected as a major point of emphasis because the 10–11-mm human can be compared with the 10-mm pig and this links the development of humans to that of other animals. The other major stage of emphasis is the 32-mm human embryo, selected because this embryo is at the end of the embryonic period and has many recognizable "adult" structures. Other stages were chosen to fill in the gaps.

This book is thus prepared with the hope that it will provide embryology students and medical students with a relatively inexpensive source of knowledge of human development to expand on the foundation they have obtained from the study of other animals.

Trent D. Stephens

Acknowledgments

I wish to especially thank Dr. Thomas H. Shepard for his help, support, and encouragement during the preparation of this book. I am grateful to him for the use of the sectioned material which comprises this work. The Central Laboratory for Human Embryology and other facilities of its kind are extremely valuable in helping to expand our knowledge of ourselves. I am grateful to all those associated with this laboratory. This research was supported in part by NIH Grant No. HD00836.

I wish to thank Dr. Ronan O'Rahilly, Director of the Carnegie Embryological Collection, for permission to reproduce the first four figures of this book from Figures 4, 8, 43, and 56 in *Developmental Stages in Human Embryos. Part A: Embryos of the First Three Weeks (Stages 1 to 9).* Carnegie Inst. of Wash., Pub. 631, Washington, D.C.

I am grateful to all of my former students who have stimulated the development of this book; and to George and Barbara Brownfield and Mary Fenucan for technical advice and manuscript preparation. Figure 37 was photographed by Paul Macapa.

Table of Contents

Introduction

Human embryos, as well as several other animal species, have been grouped into a series of developmental stages to facilitate comparison of developmental states and gestational ages between embryos. Streeter (1951, *Developmental Horizons in Human Embryos,* Carnegie Inst. of Washington Pub.) devised 23 developmental horizons to classify human embryos from fertilization to the approximate end of the embryonic period (60 days). In recent years the term "developmental horizon" has been replaced by most investigators with the term "developmental stage" and the Roman numerals introduced by Streeter have been replaced by Arabic.

These developmental stages will be referred to throughout the text. Unfortunately, the human developmental stages do not correspond to similar developmental stages of other animals, but comparisons can be made. The following chart describes the major features of the 23 developmental stages of human embryos including gestational age and crown-rump length.

Human Developmental Stages*

Stage	Length (mm)	Age (Days)**	Somite Pairs	Characteristics
1	0.1	0–1		Fertilized uncleaved zygote
2	0.1–0.2	2–3		segmentation = two cells to morula
3	0.1–0.2	4–5		Unimplanted, free-floating blastocyst
4	0.2–0.3	6–7		Implantation
5	0.3–0.5	8–11		Progression into endometrium
6	0.2–0.5***	12–14		Embryonic disc, Villi, and yolk sac appear
7	0.3–0.7	15–16		Primitive streak appears
8	0.5–2.0	17–18		Neural folds elevate
9	1.5–3.0	19–20	1–3	Head fold appears
10	2.0–3.0	21–23	4–12	Neural fold fusion begins, heart begins to beat
11	2.5–3.0	23–25	13–20	Two branchial arches, foregut, hindgut, optic evagination
12	3.0–4.0	21–29	21–29	Arm buds appear, neural tube closed, optic cup
13	4–5	28–30	40 (complete no.)	Leg buds appear, heart chambers, lung buds, metanephric bud
14	6–7	30–32		Lens invagination, septum primum, gonadal ridge
15	7–8	32–34		Lens vesicle closed, external ears becoming recognizable
16	9–10	35–36		Eye pigment appears, hand plate, hypophysis, liver
17	11–14	37–40		Finger rays, foot plate, ear defined, somites less apparent superficially
18	14–16	40–42		Eyelid, finger rays notched, toe rays, nerve plexuses
19	17–20	42–44		Head more erect, limbs extend forward, muscles developing, duodenum closed
20	21–23	45–46		Fingers, scalp plexus present, optic nerve, septum secundum
21	22–24	46–48		Hands meet over heart region, corpus striatum, thalamus, heart valves
22	25–27	48–50		Fingers overlap those of opposite hand, duodenum reopened
23	28–30	50–52		Head erect and rounded, scalp plexus reaching head vertex, ossification begins

*From Streeter GL (1959): *Developmental Horizons in Human Embryos.* Carnegie Institution of Washington, Washington, D.C.; O'Rahilly R. (1973): *Developmental Stages in Human Embryos. Part A: Embroys of the First Three Weeks (Stages 1 to 9).* Carnegie Institution of Washington, Pub. 631, Washington, D.C.

**Gestational age in days established from Iffy L, Shepard TH, Jakobovits A, Lemire RJ, and Kerner P (1967): The rate of growth in young human embryos of Streeter's horizon XIII to XXIII. *Acta Anat.* 66:178–186.

***Stage six refers to embryonic disc only, stage 5 refers to overall size. Stages thereafter refer to crown-rump length.

Early Human Embryos

Early Human Embryos

The first four figures* were chosen to represent some of the early stages of human embryology.

Figure 1, a two-celled embryo, is representative of stage 2. During this stage several cell divisions occur taking the embryo from two to perhaps twenty cells.

Figure 2, a 107-celled blastocyst, is characteristic of stage 3 and is 4–5 days old. This stage has developed a blastocoele but has not yet begun to implant.

Figure 3 is an embryo 15 days (2 weeks) old and is at developmental stage 7. It consists of an embryonic disc measuring 0.3–0.7 mm in diameter. The embryo has implanted into the endometrial lining of the uterus and has developed villi, a yolk sac, and an amnion. The embryo at this stage is bilaminar and is developing a primitive streak.

Figure 4 is an embryo 17 days (2.5 weeks) old and is at developmental stage 8. The embryo at this stage is trilaminar, has a well-defined primitive streak and Henson's node, and has the beginning of neural folds.

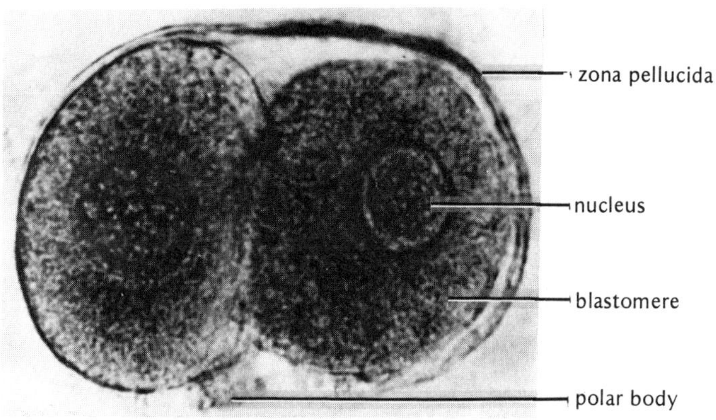

Figure 1 Human embryo after first sell division, two-celled embryo

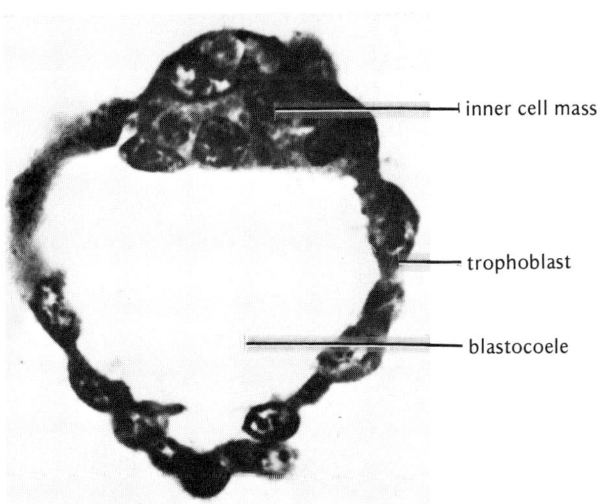

Figure 2 Human embryo at the blastocyst stage

*Reproduced by permission from O'Rahilly R. (1973): *Developmental Stages in Human Embryos. Part A: Embryos of First Three Weeks (Stages 1 to 9).* Carnegie Institution of Washington. Publication 631. Figures 4, 8, 43, and 56.

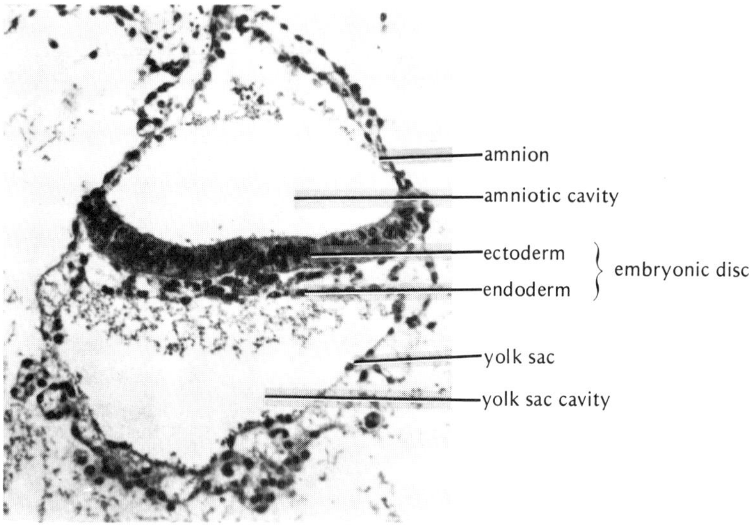

Figure 3 Human embryo at embryonic disc stage

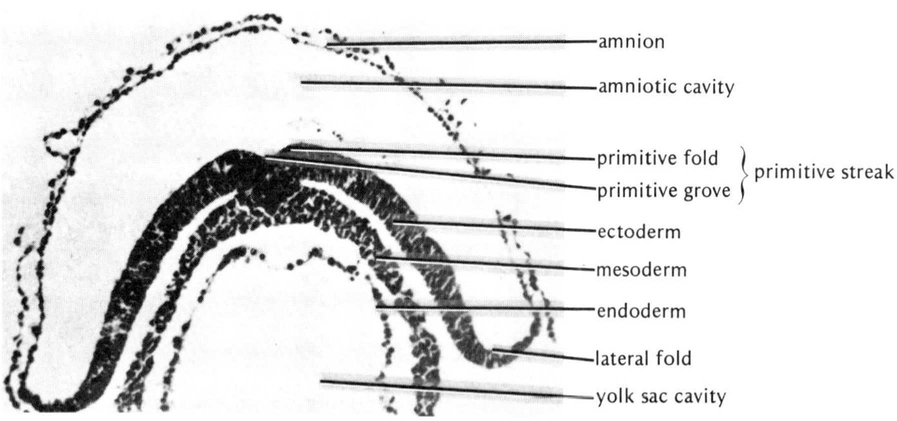

Figure 4 Human embryo at primitive streak stage

6.5-mm Human Embryo

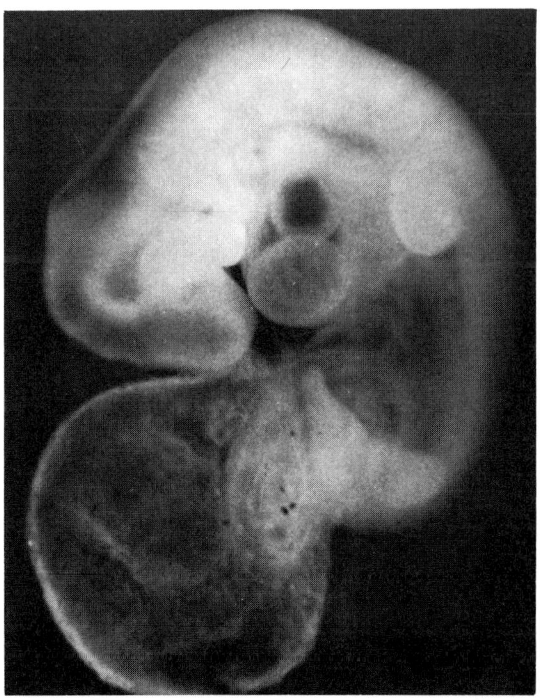

Figure 5 6.5-mm human embryo

6.5-mm Human Embryo

The specimen chosen to represent this stage is H-1099, a male. The embryo was preserved in Lillies fixative, sectioned at 10μ, and stained with PAS. The 6.5-mm human embryo is approximately 32 days (4 weeks) old and is at developmental stage 14.

Following is a list of some of the changes which have occurred since stage 8. The neural folds have elevated and closed into a tube. All 40 somites have appeared. The limb buds have appeared. Three branchial arches are present. The optic cup has evaginated from the brain and the lens vesicle is invaginating. Otic and olfactory placodes have appeared and the former has become a vesicle. The heart has fused in the midline, begun to beat, and has become divided into four primitive chambers. The foregut and hindgut have developed; the stomatodeum has ruptured; and the primordia of the thyroid, liver, pancreas, and lungs can be seen associated with the developing gastrointestinal tube. The mesonephros has developed and the metanephros is appearing.

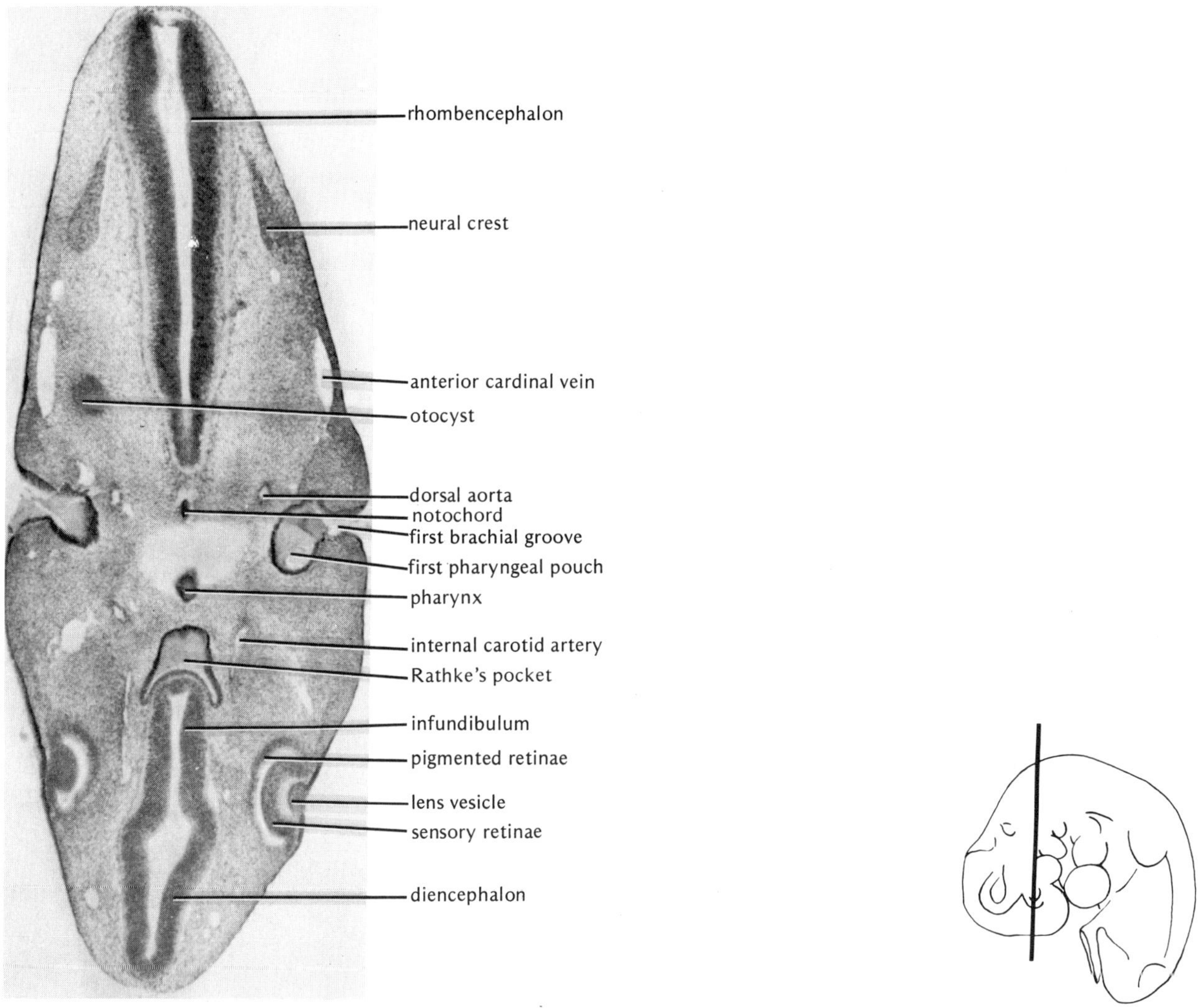

Figure 6 6.5-mm human embryo, transverse section at brain level

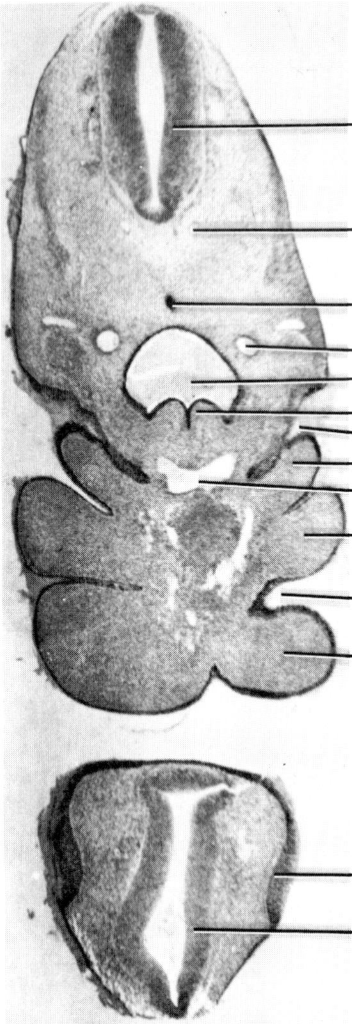

Figure 7 6.5-mm human embryo, transverse section at pharynx level

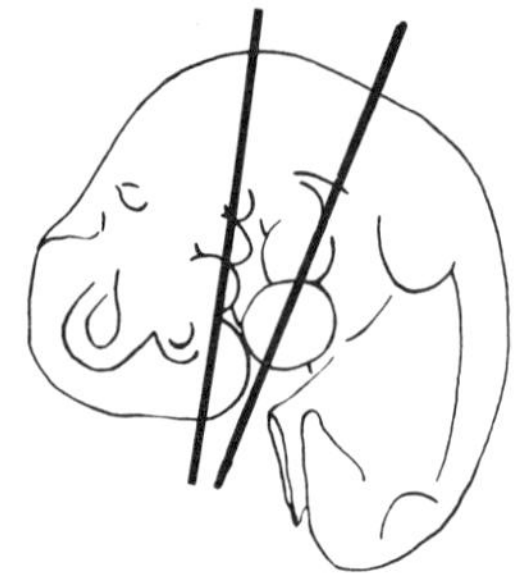

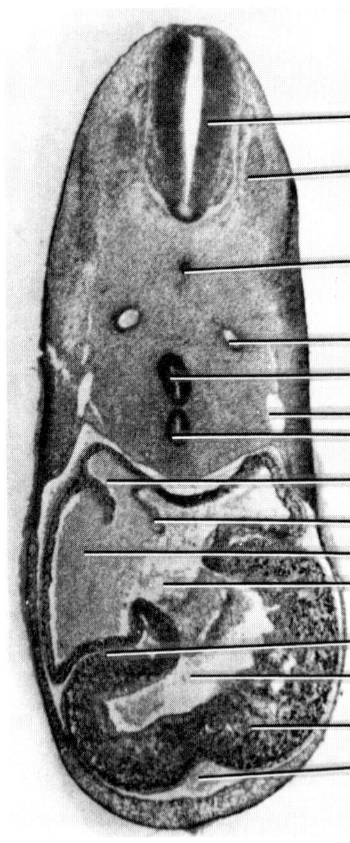

Figure 8 6.5-mm human embryo, transverse section at heart level

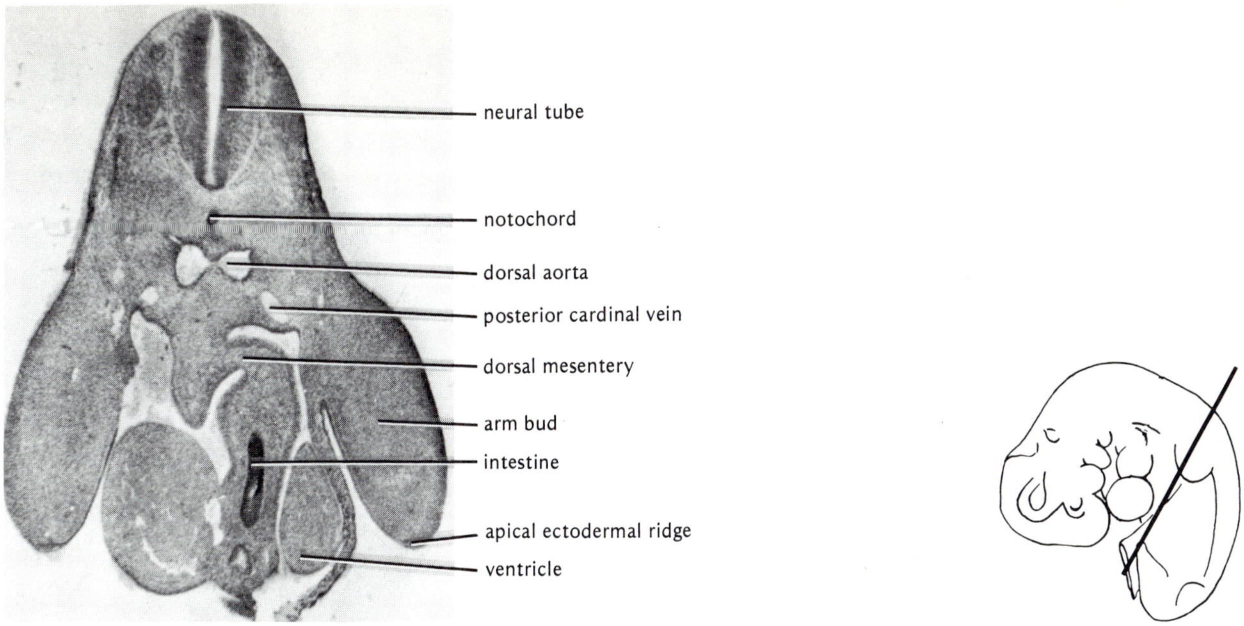

Figure 9 6.5-mm human embryo, transverse section at arm level

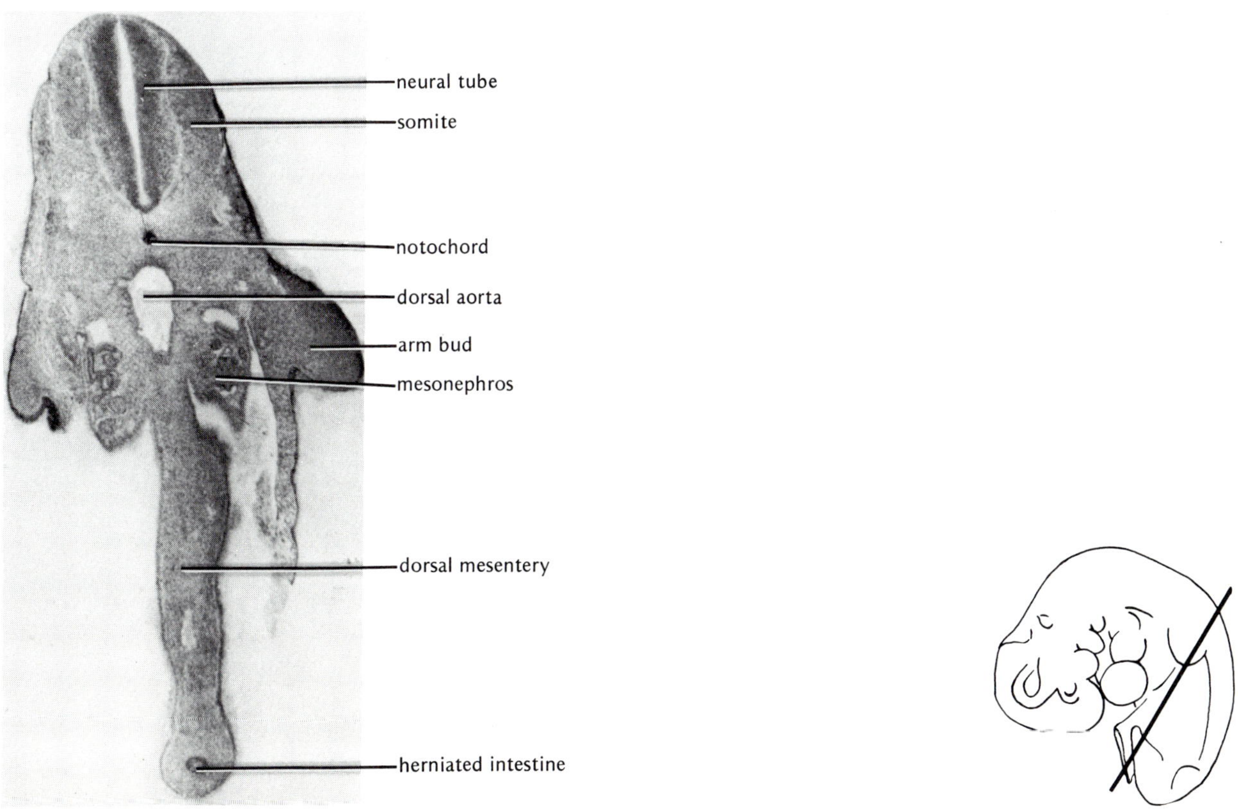

Figure 10 6.5-mm human embryo, transverse section at mesonephros level

11.5-mm Human Embryo

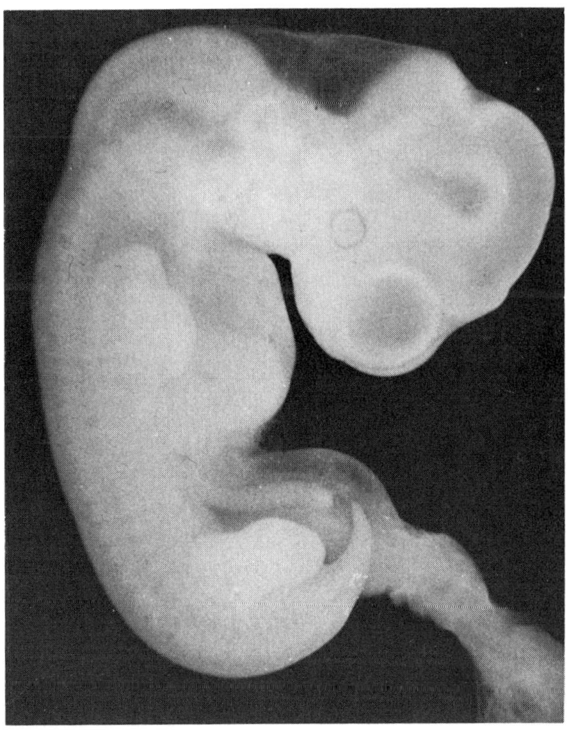

Figure 11 11.5-mm human embryo

11.5-mm Human Embryo

The specimen chosen to represent this stage is H-1870, a male. The embryo was preserved in Lillies fixative, sectioned at 10μ, and stained with PAS. The 11.5-mm human embryo is approximately 35 days (5 weeks) old and is at developmental stage 17.

Following is a list of some of the changes which have occurred since stage 14 (6.5 mm). The limbs have developed distal plates and the hand plate exhibits digital rays. The lens vesicle has closed off from the surface and eye pigment has appeared. The cerebral hemispheres and cerebellum are developing. The hypophysis is developing. The nasal swellings are developing. Nerve plexuses are forming. The septum primum and interventricular muscular septum are developing. The gonadal ridge and Müllerian ducts appear. The lung lobes have developed and the liver is present. The intestinal hernia has developed and the stomach has begun to rotate.

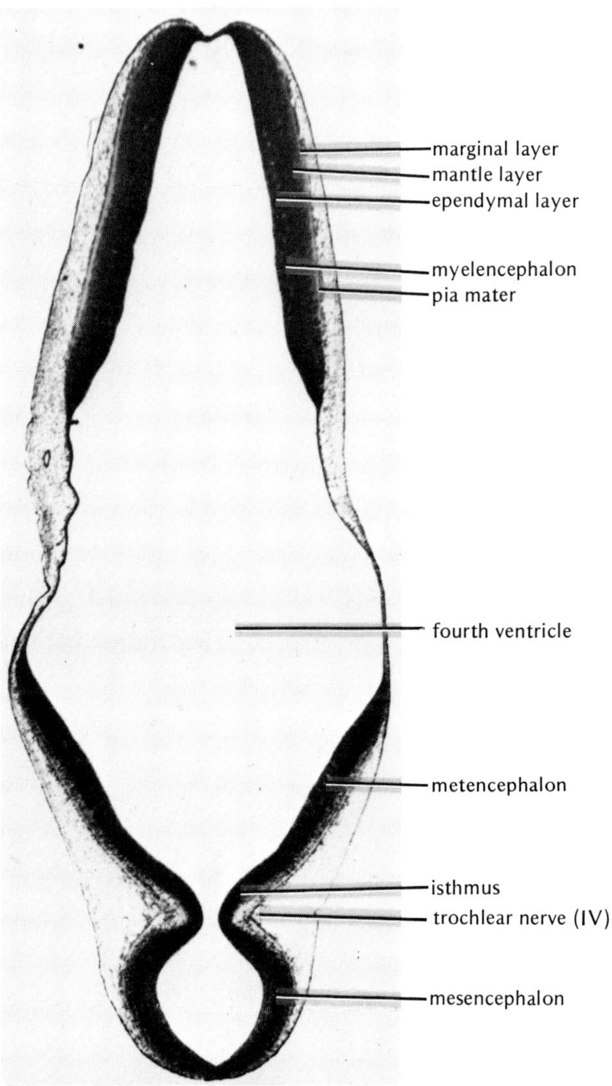

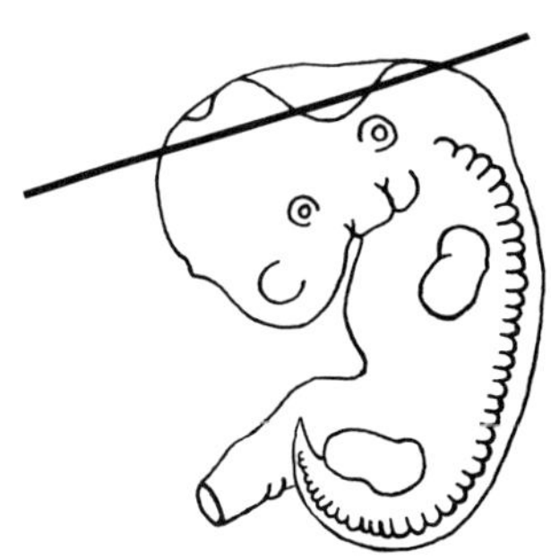

Figure 12 11.5-mm human embryo, transverse section at rhombencephalon level

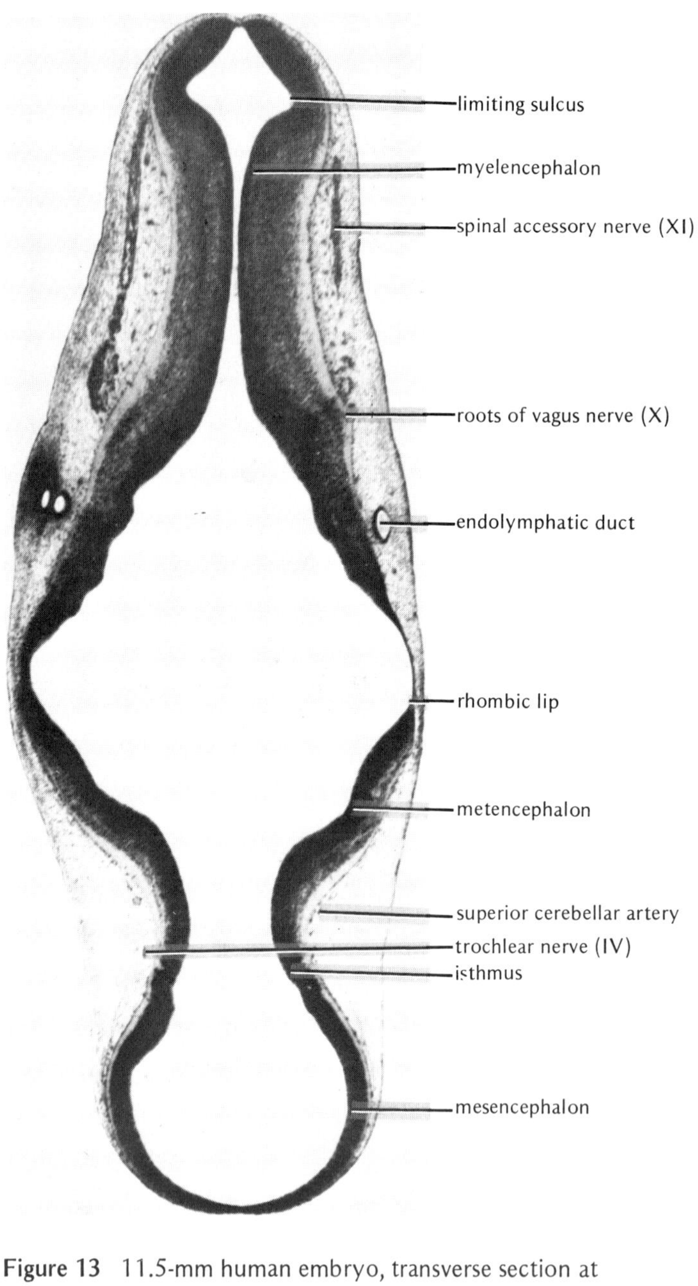

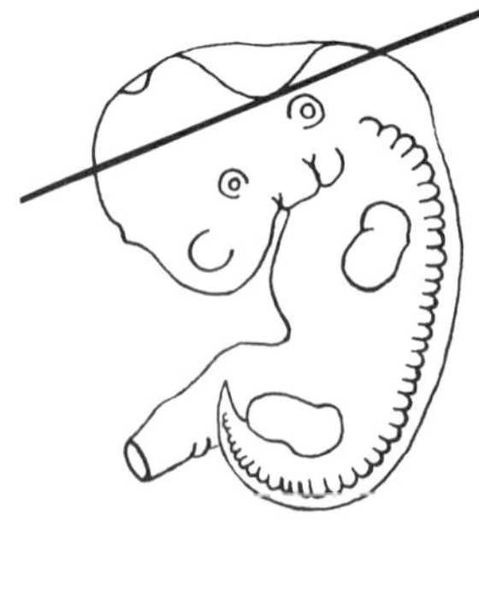

Figure 13 11.5-mm human embryo, transverse section at endolymphatic duct level

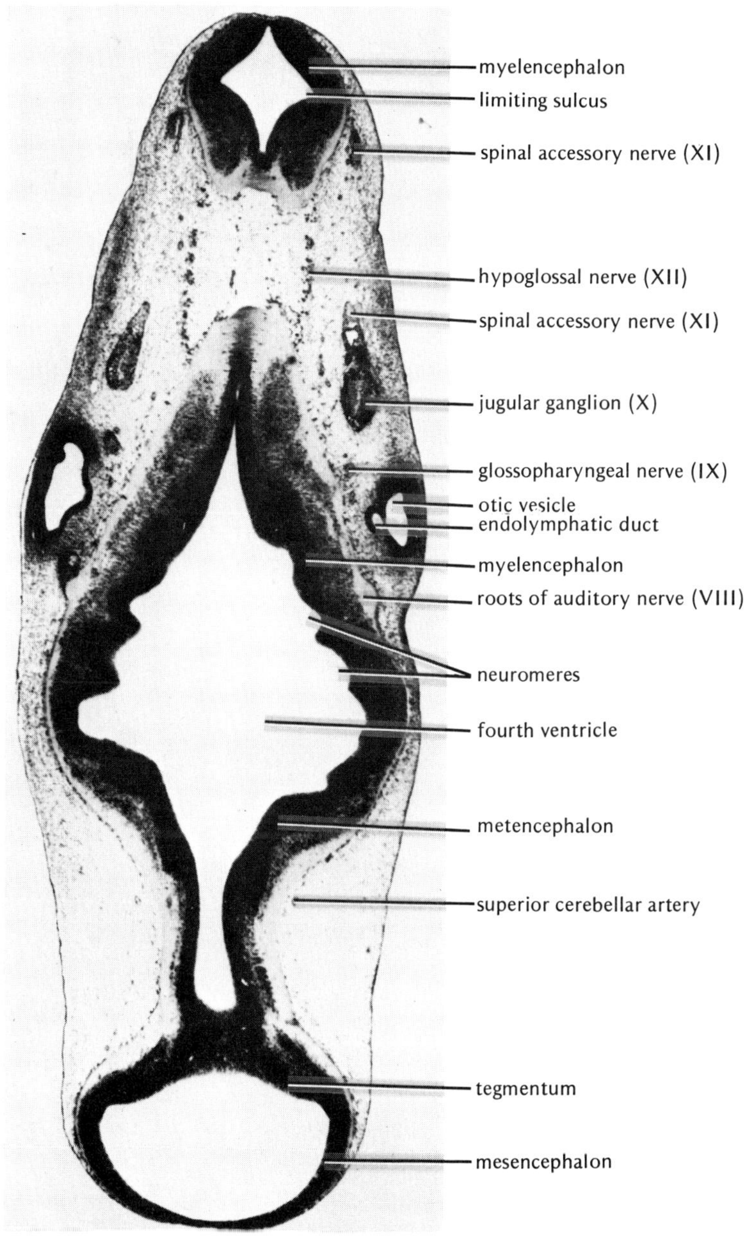

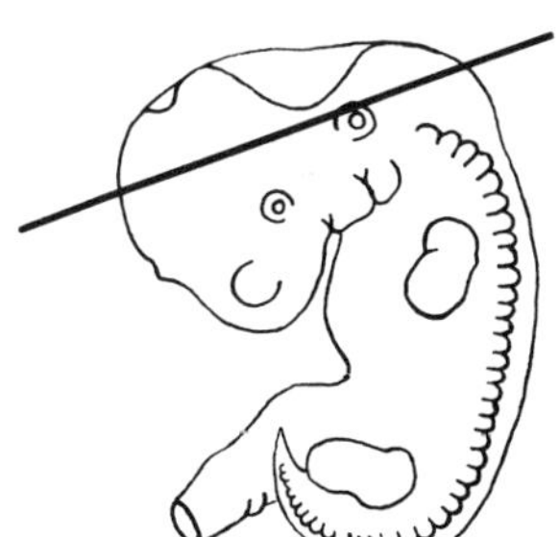

Figure 14 11.5-mm human embryo, transverse section at inner ear level

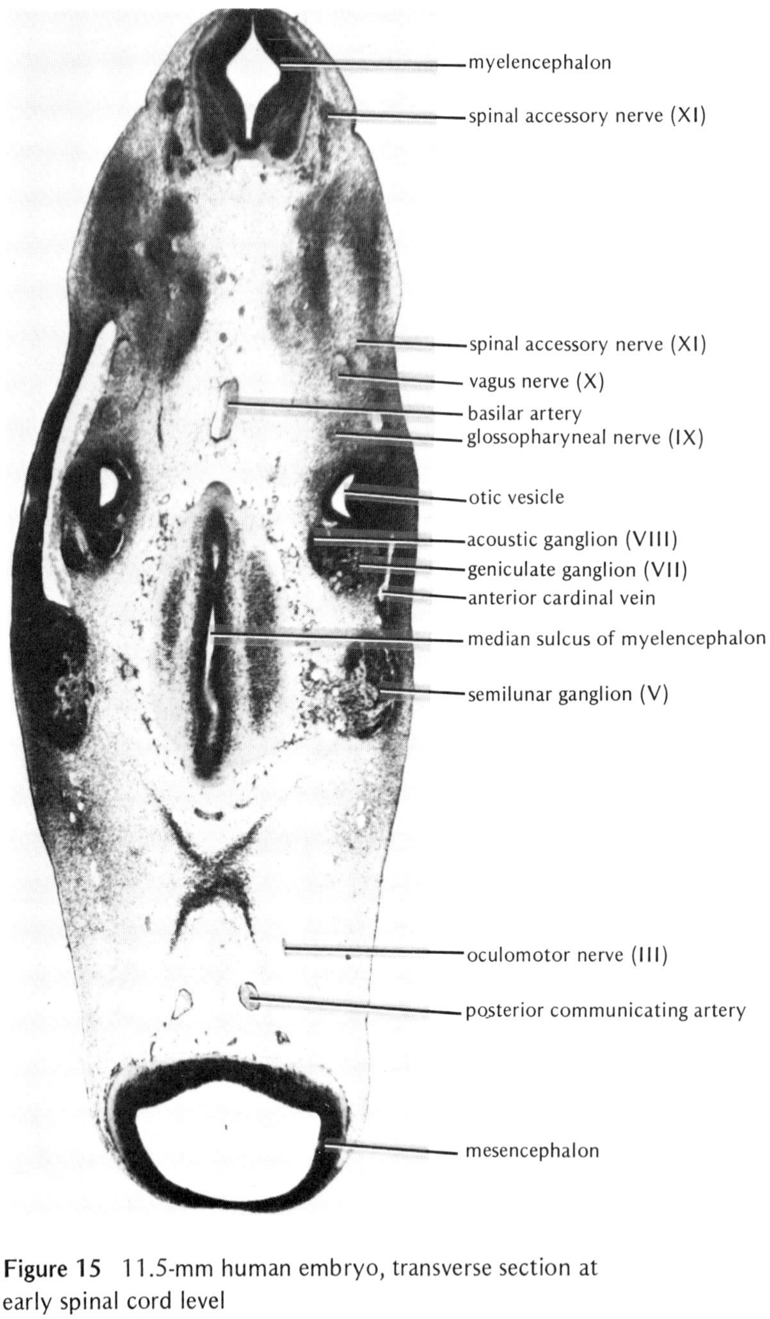

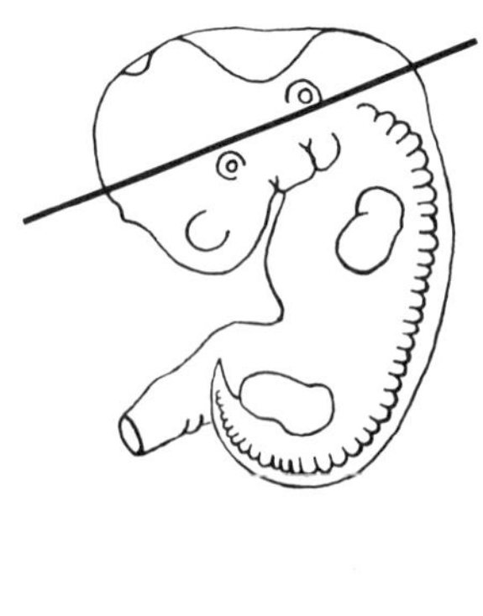

Figure 15 11.5-mm human embryo, transverse section at early spinal cord level

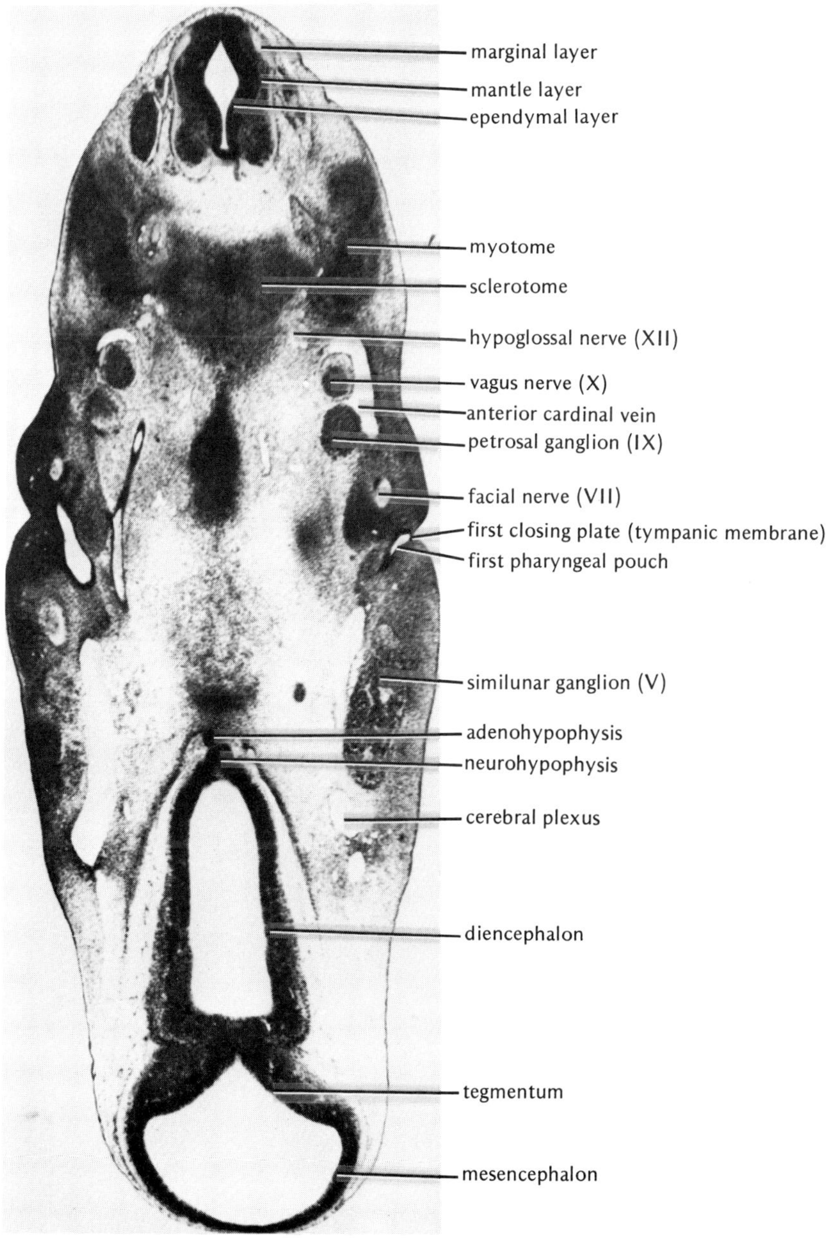

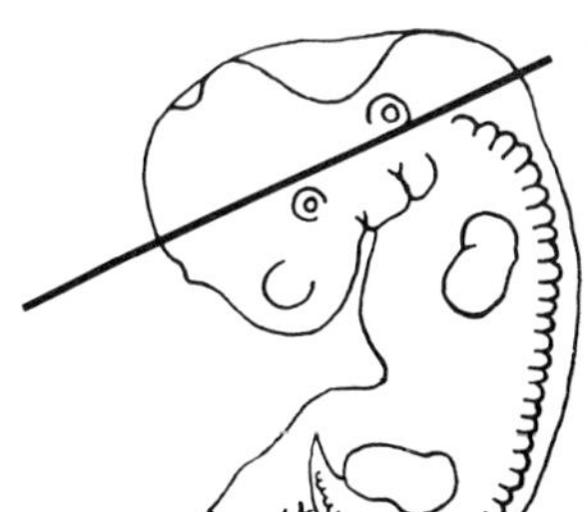

Figure 16 11.5-mm human embryo, transverse section at first pharyngeal pouch level

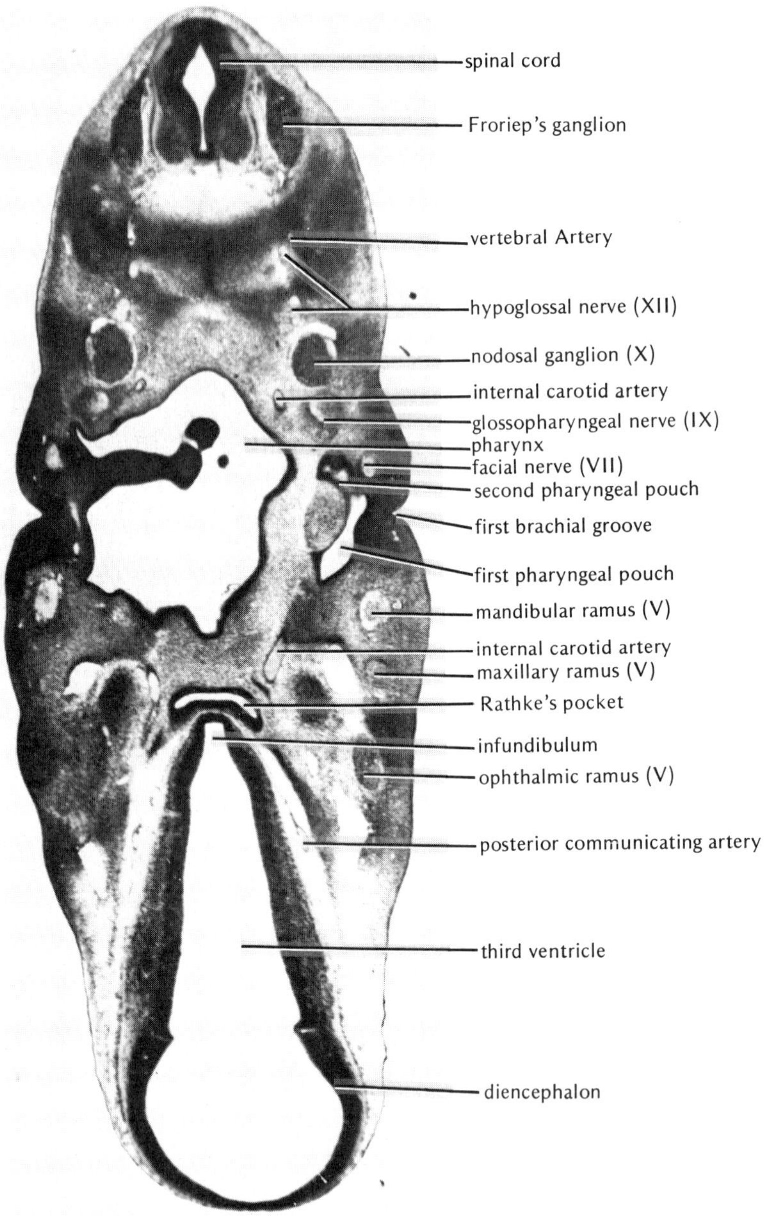

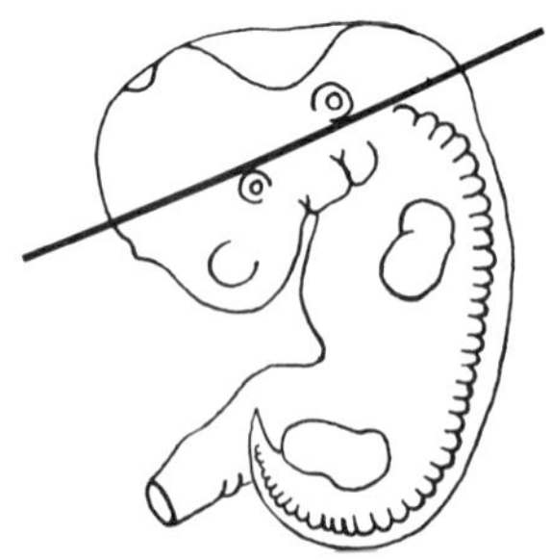

Figure 17 11.5-mm human embryo, transverse section at pharynx level

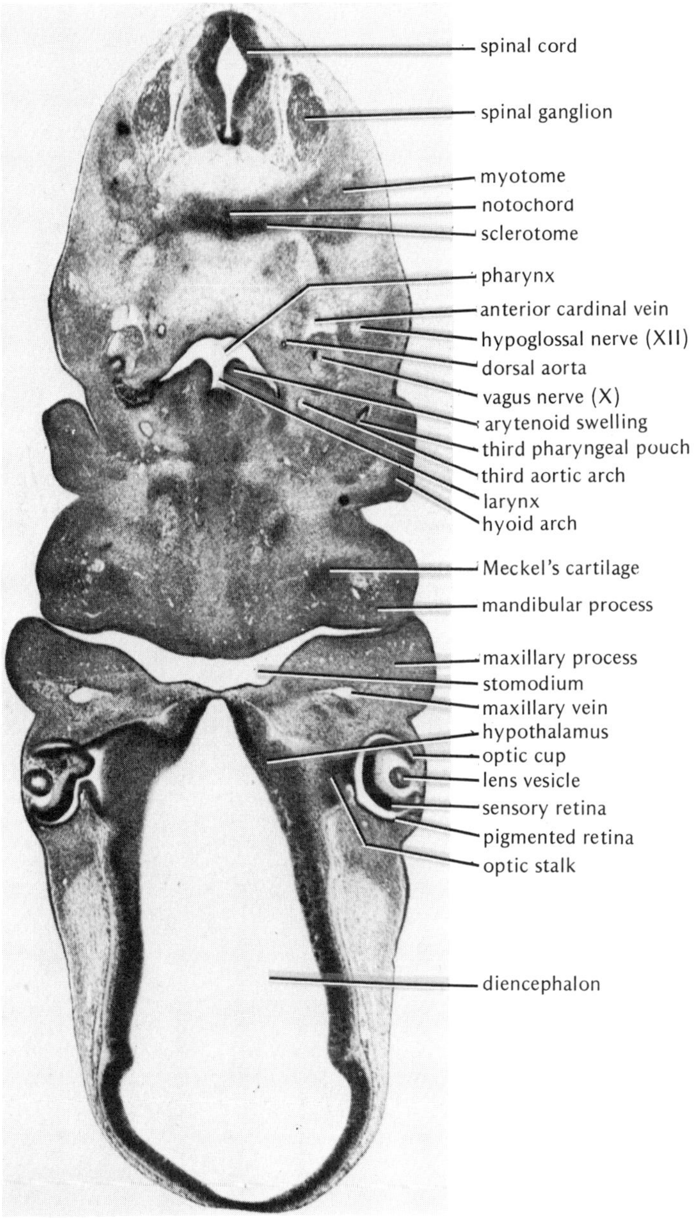

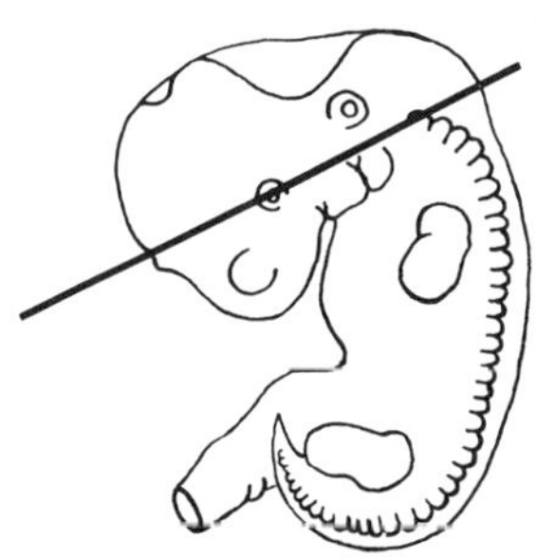

Figure 18 11.5-mm human embryo, transverse section at eye level

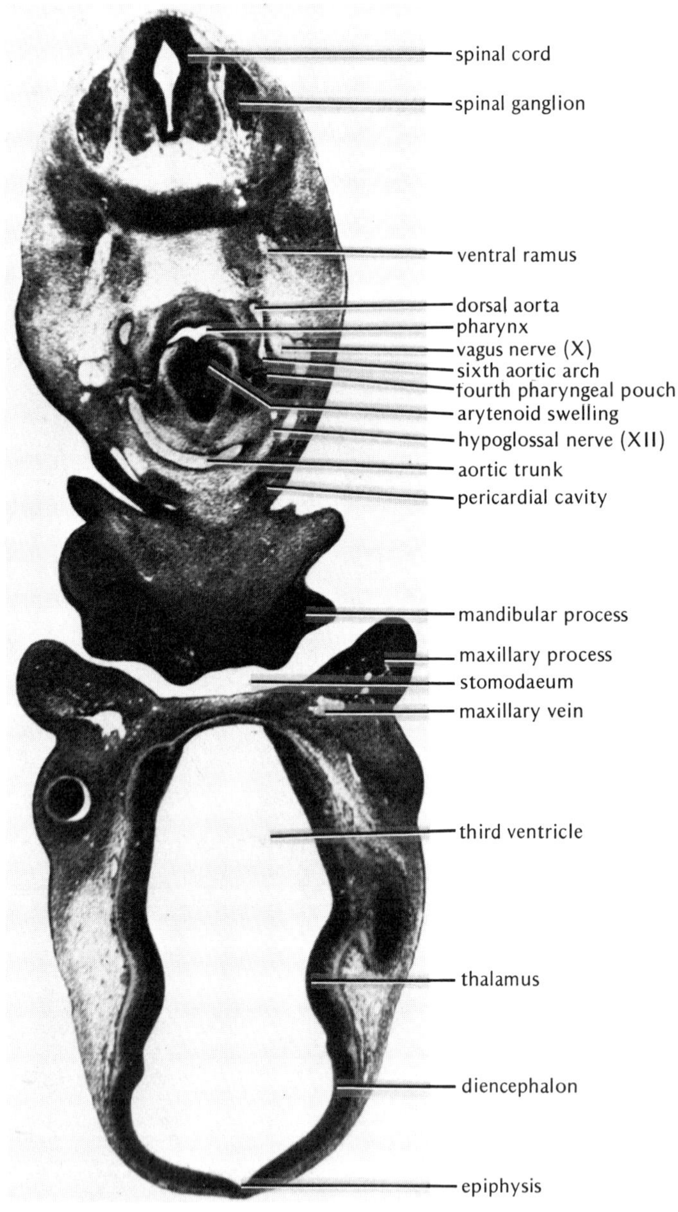

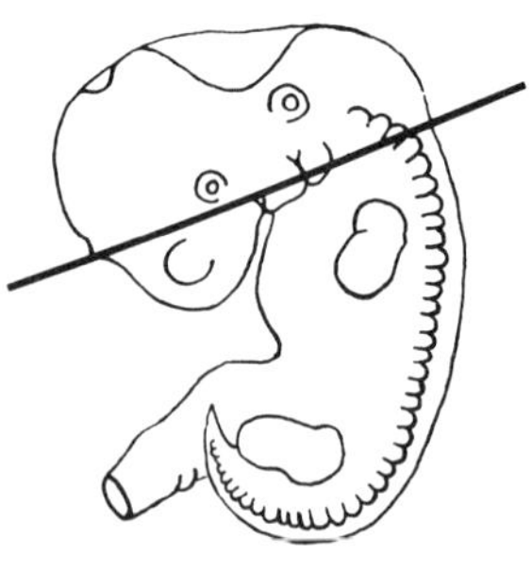

Figure 19 11.5-mm human embryo, transverse section at forebrain level

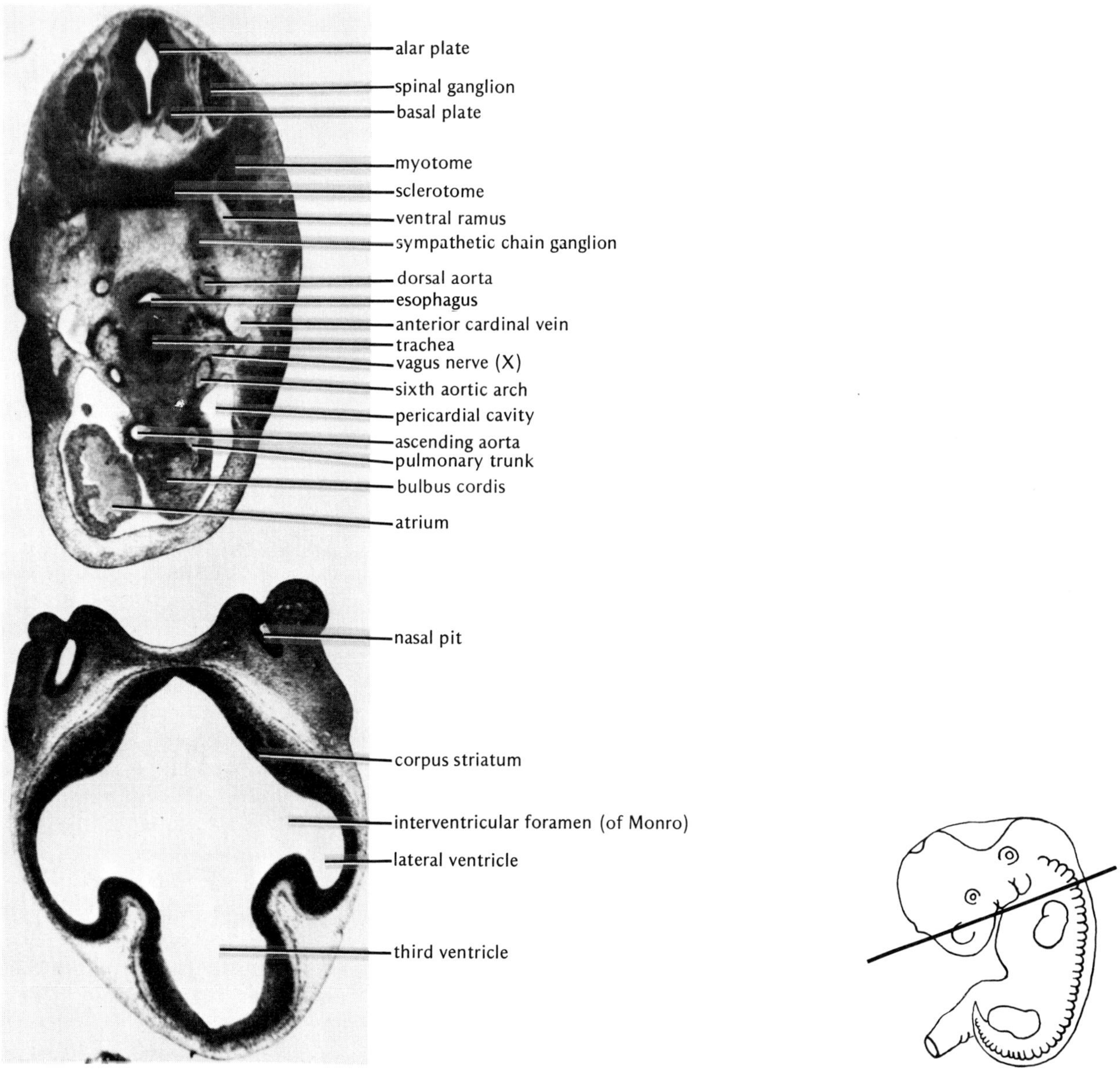

Figure 20 11.5-mm human embryo, transverse section at ventricular foramen level

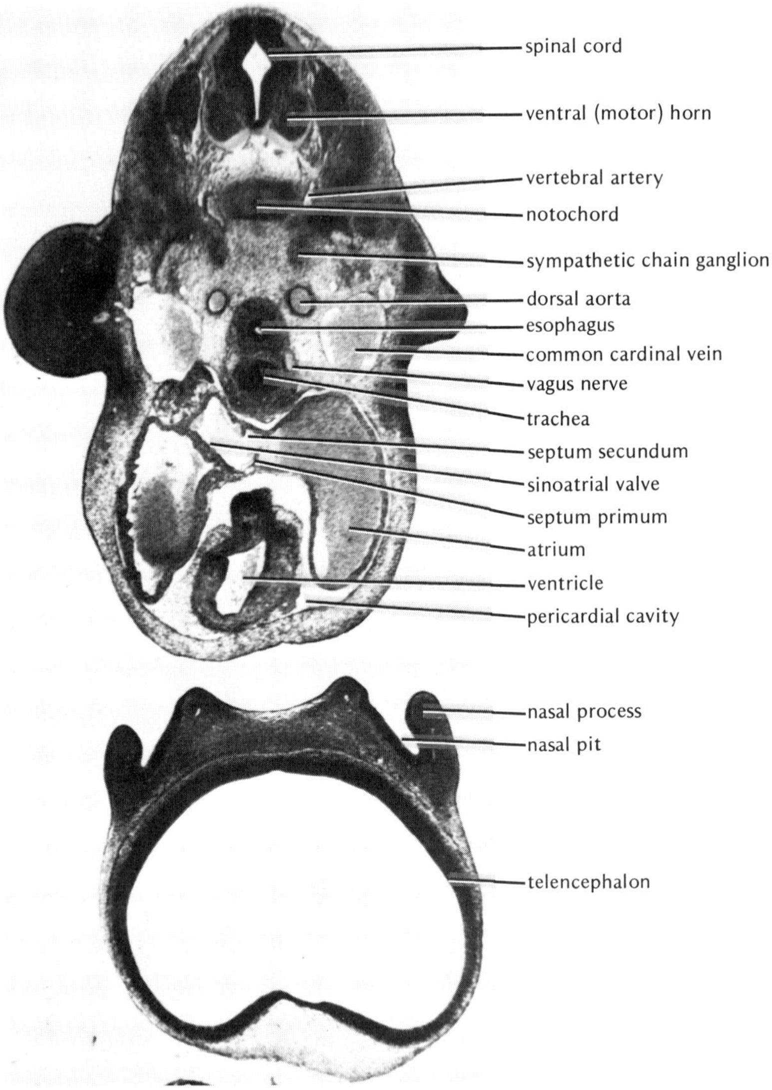

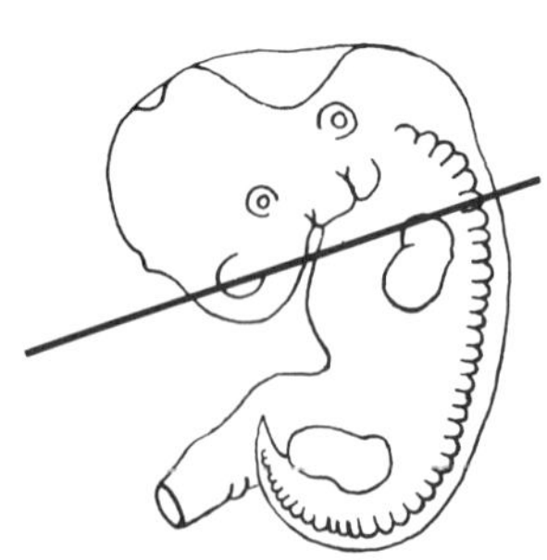

Figure 21 11.5-mm human embryo, transverse section at external nares level

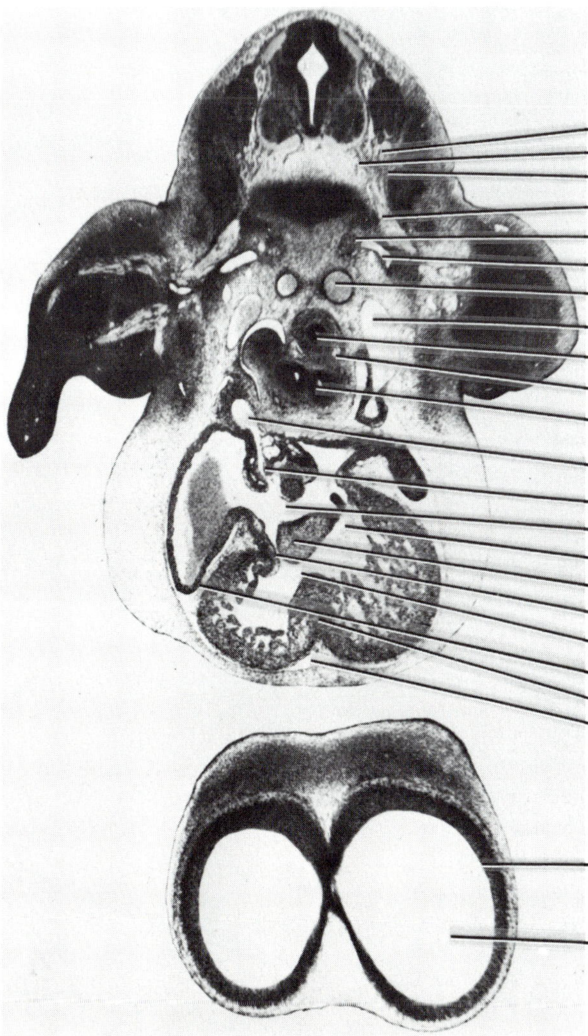

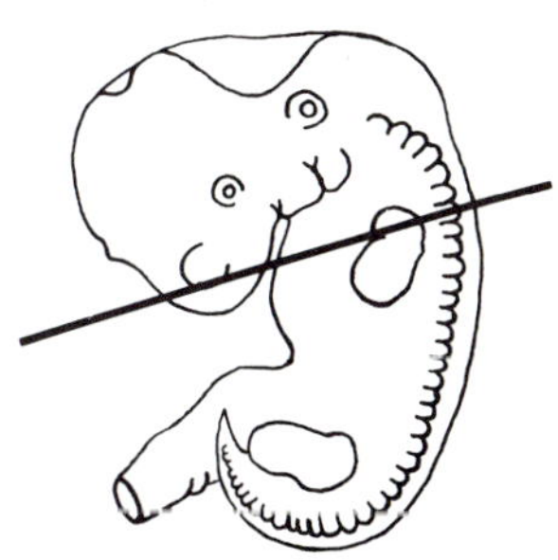

Figure 22 11.5-mm human embryo, transverse section at heart level

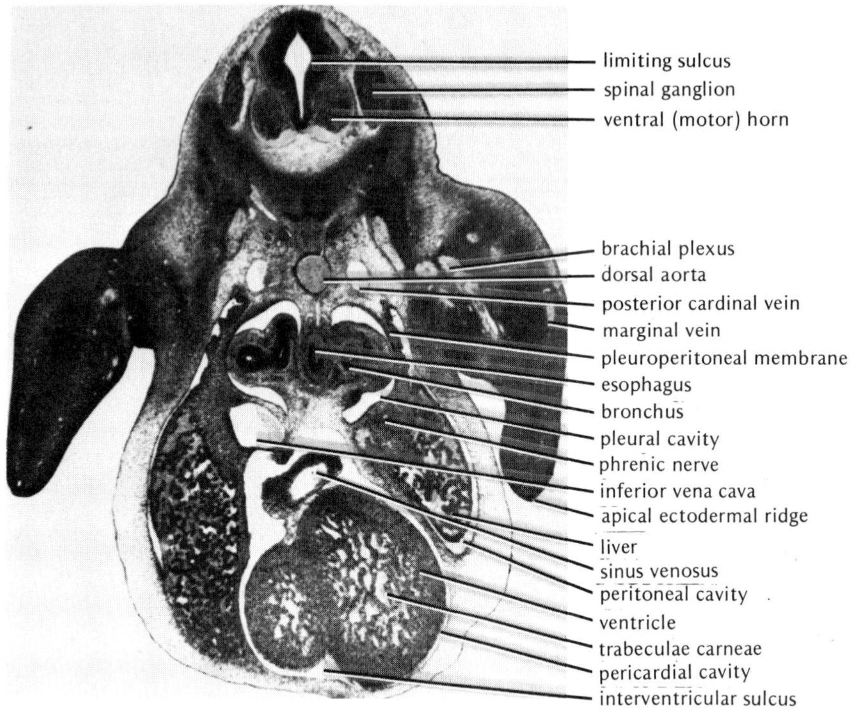

Figure 23 11.5-mm human embryo, transverse section at arm level

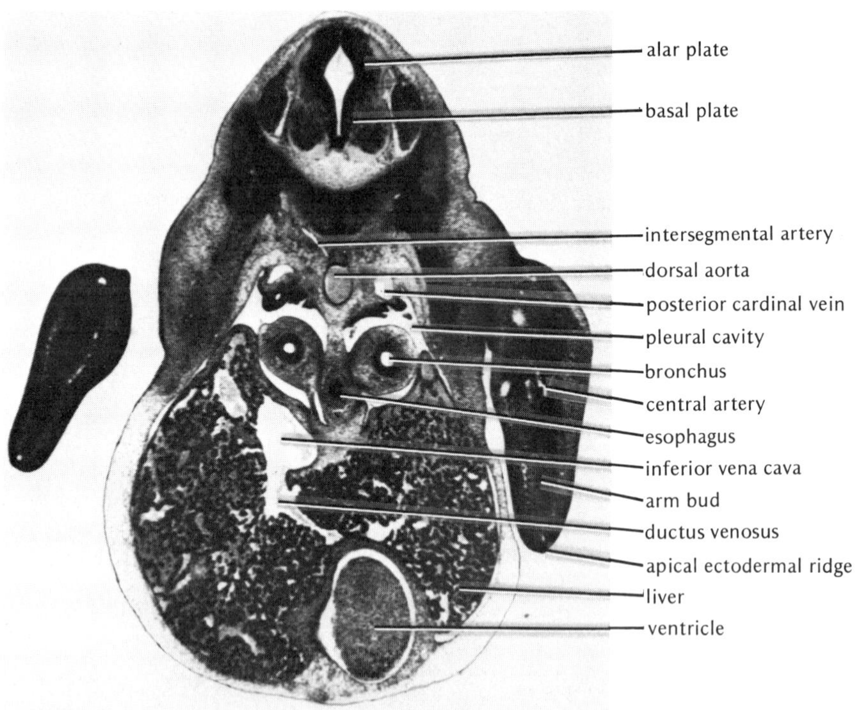

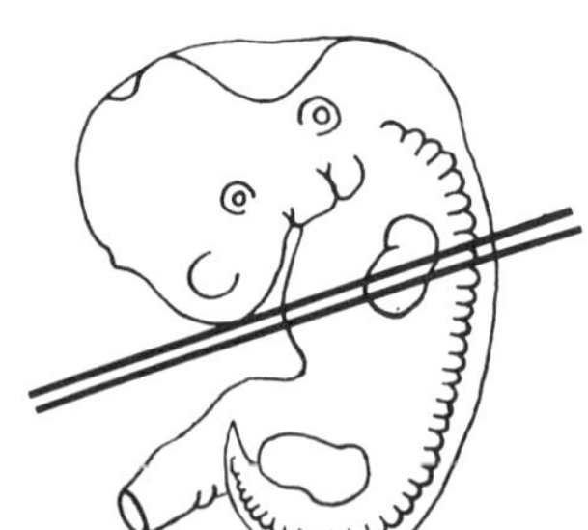

Figure 24 11.5-mm human embryo, transverse section at lung level

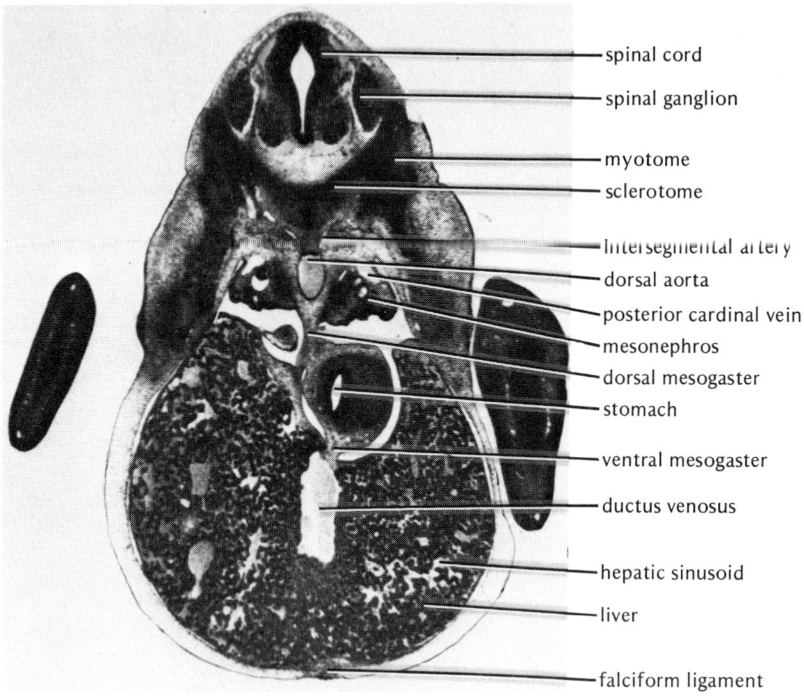

Figure 25 11.5-mm human embryo, transverse section at liver level

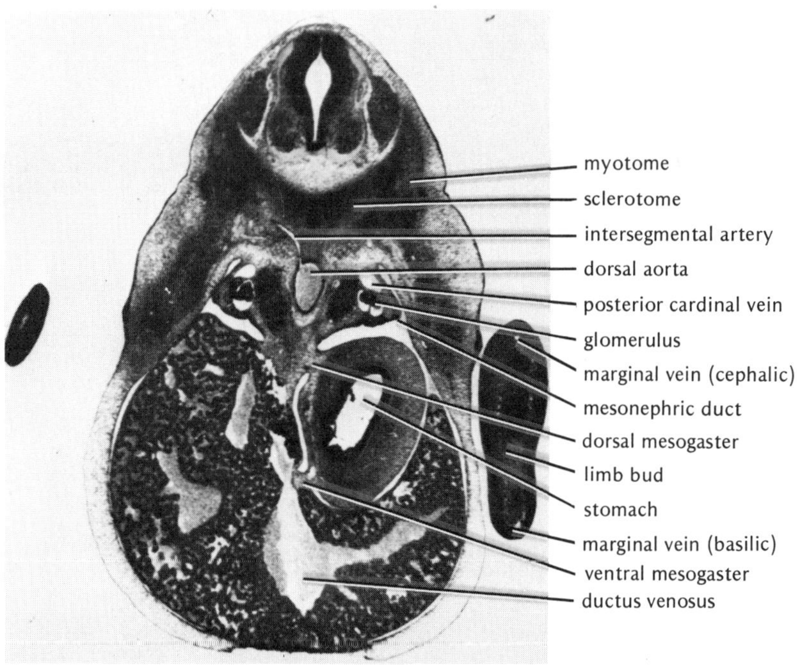

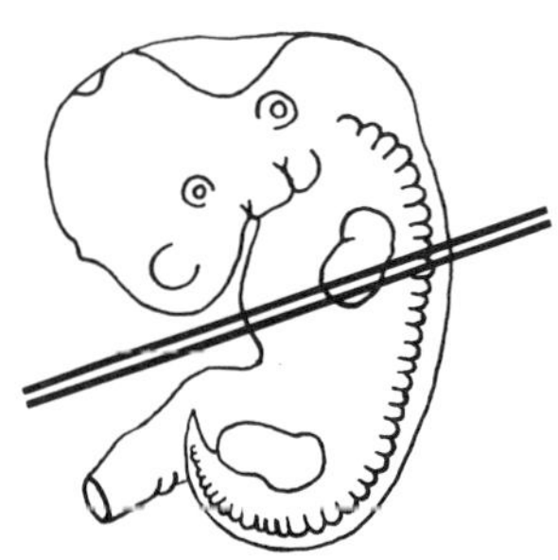

Figure 26 11.5-mm human embryo, transverse section at stomach level

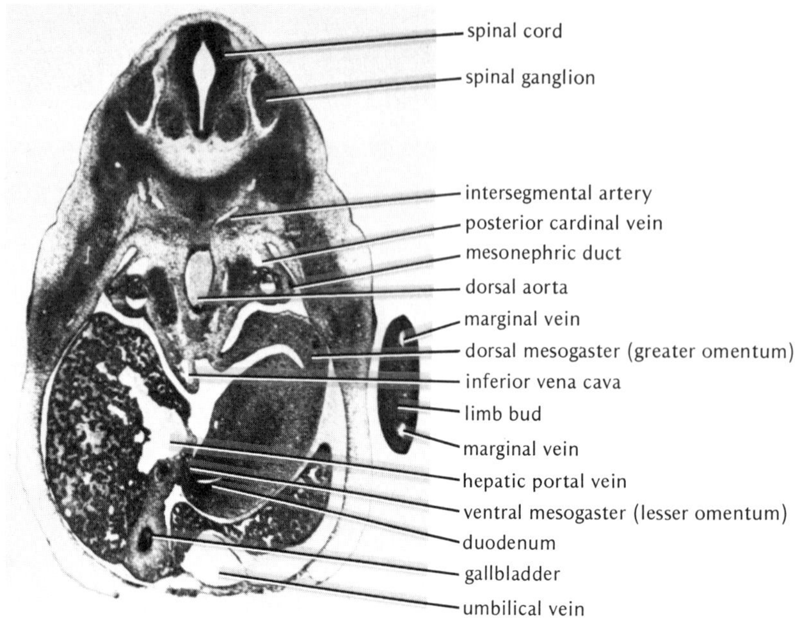

Figure 27 11.5-mm human embryo, transverse section at duodenum level

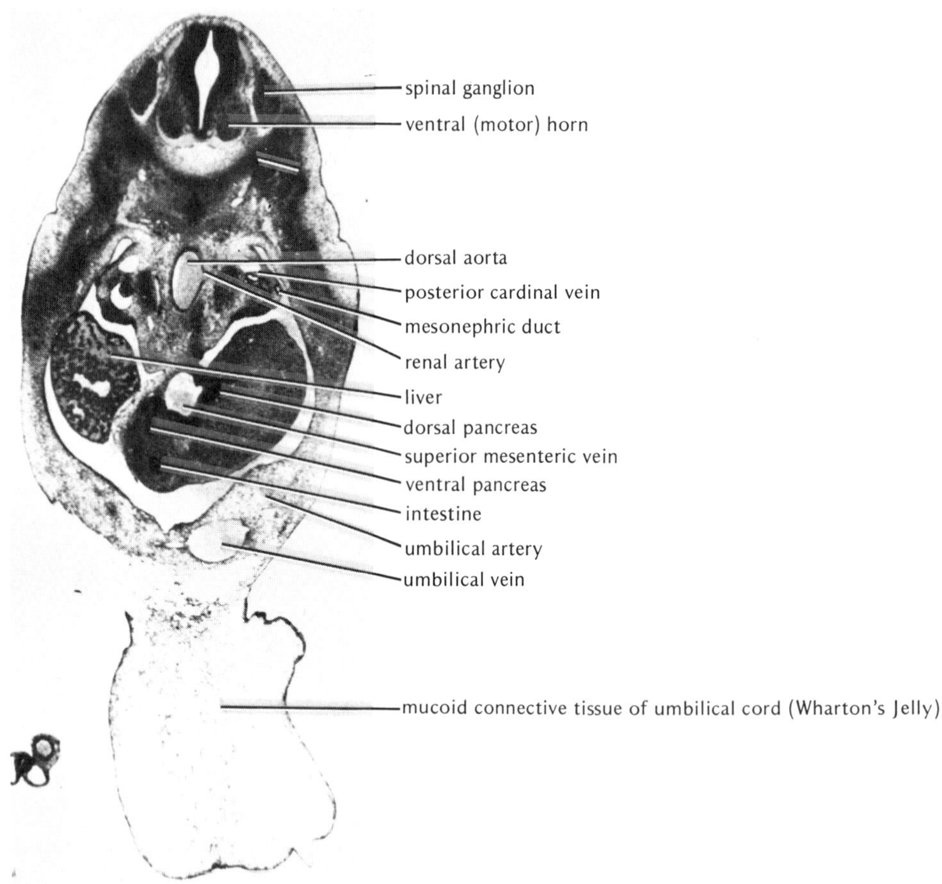

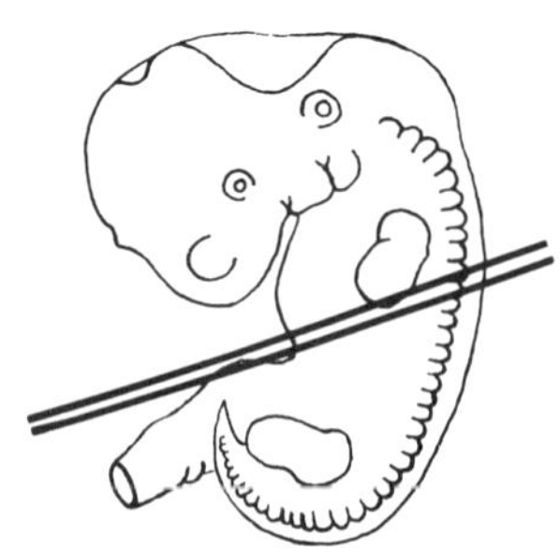

Figure 28 11.5-mm human embryo, transverse section at pancreas level

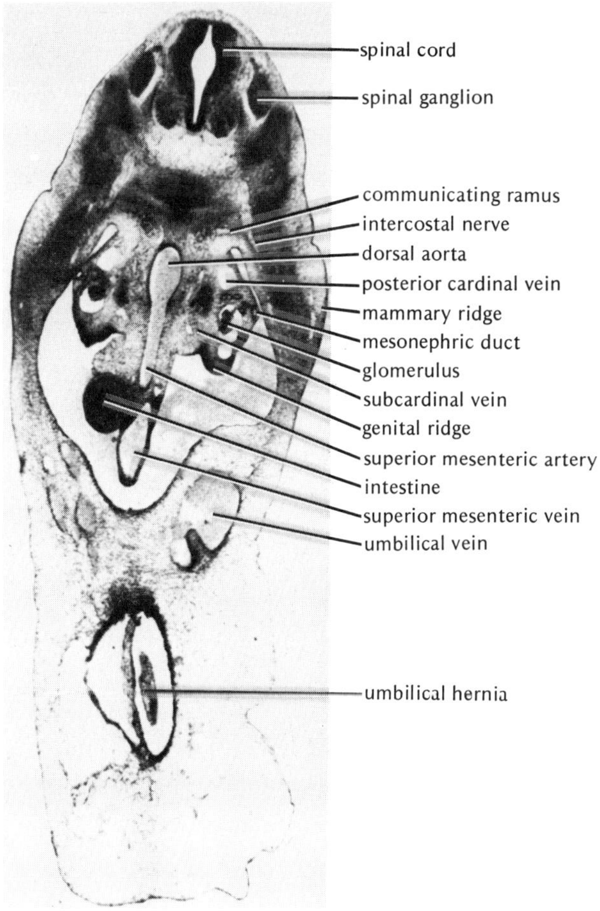

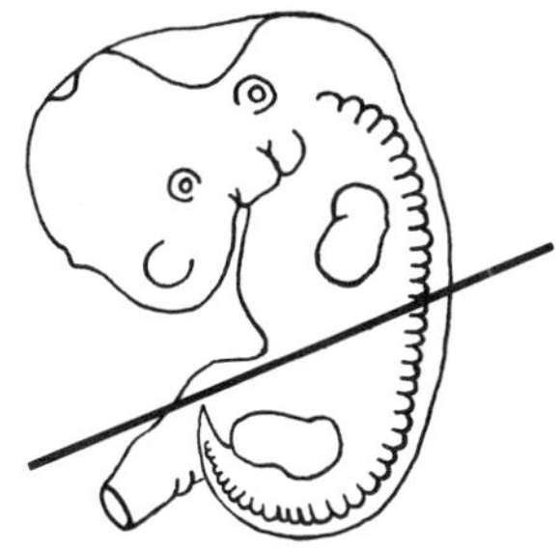

Figure 29 11.5-mm human embryo, transverse section at superior mesonteric vein level

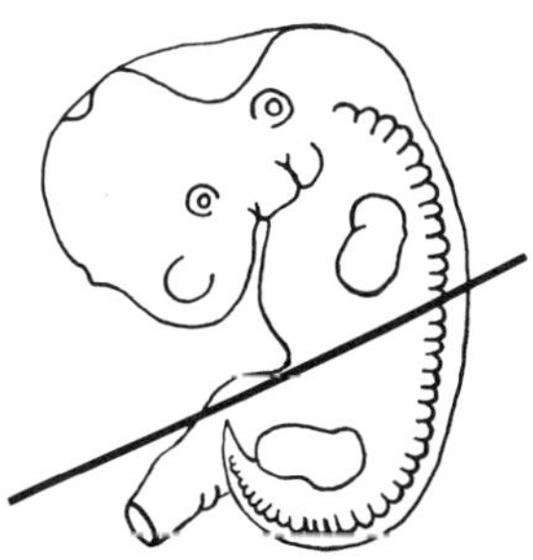

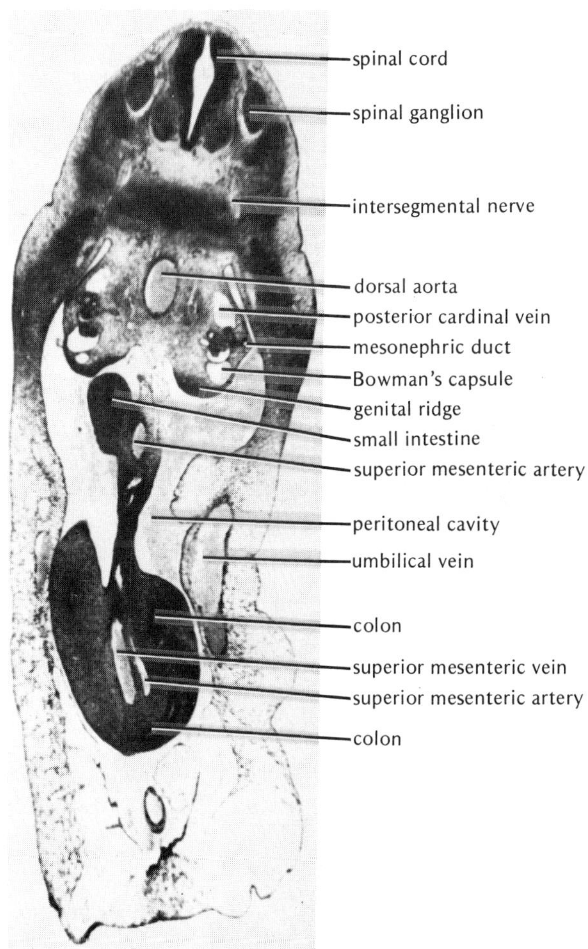

Figure 30 11.5-mm human embryo, transverse section at umbilical hernia level

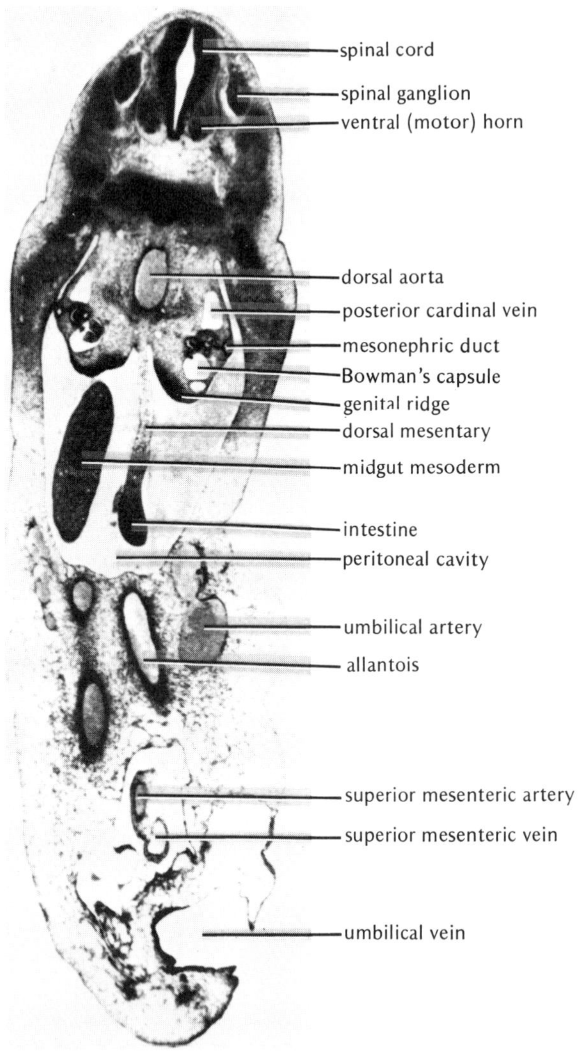

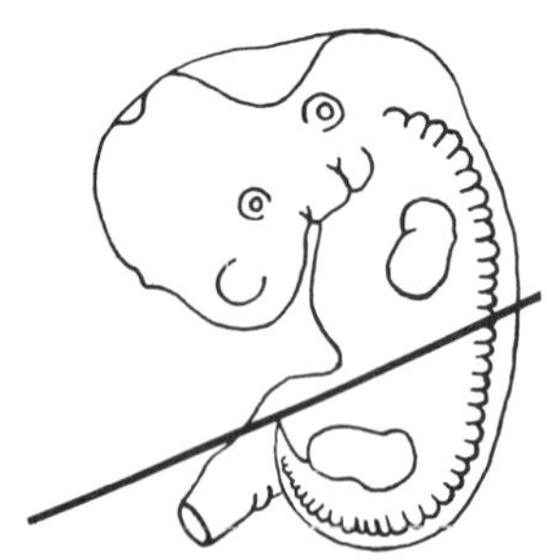

Figure 31 11.5-mm human embryo, transverse section at gonad level

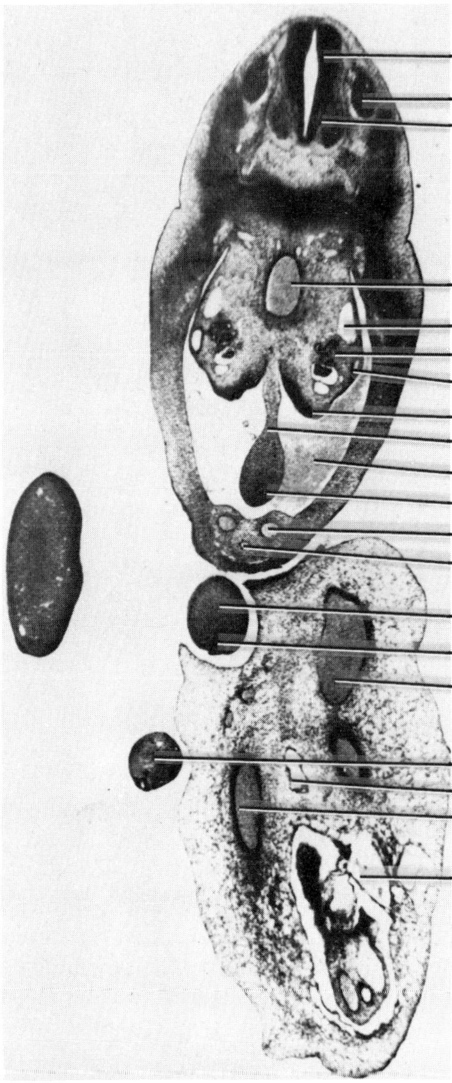

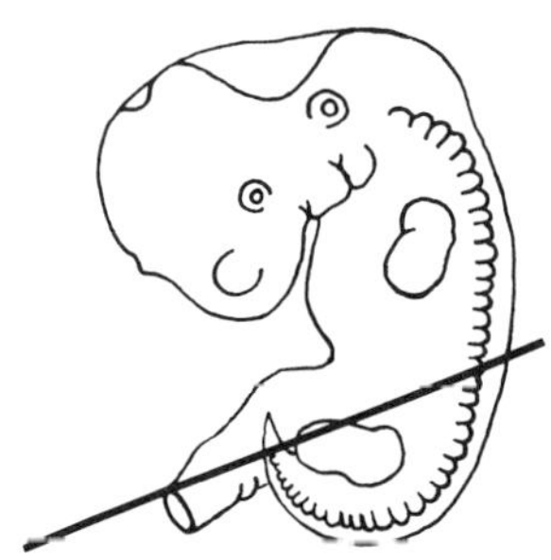

Figure 32 11.5-mm human embryo, transverse section at mesonephros level

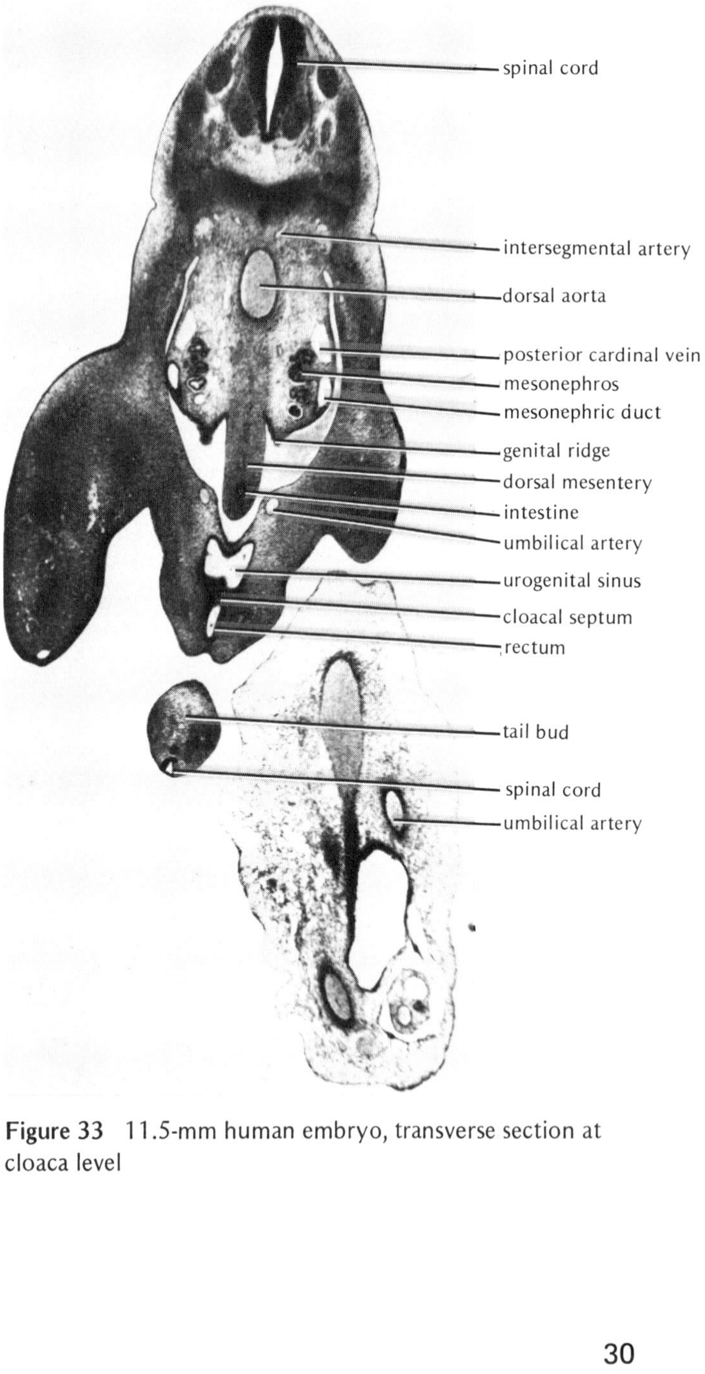

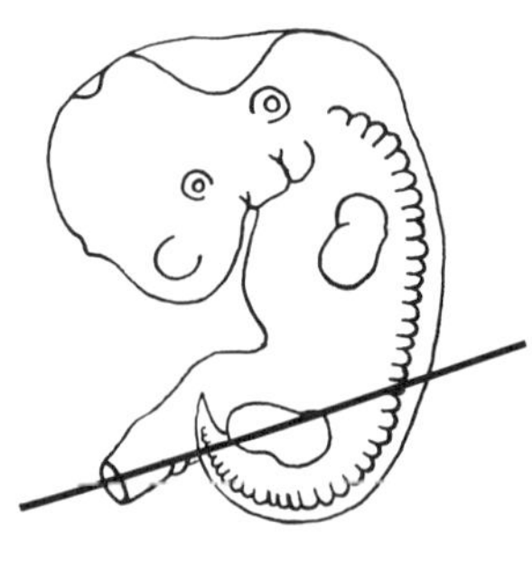

Figure 33 11.5-mm human embryo, transverse section at cloaca level

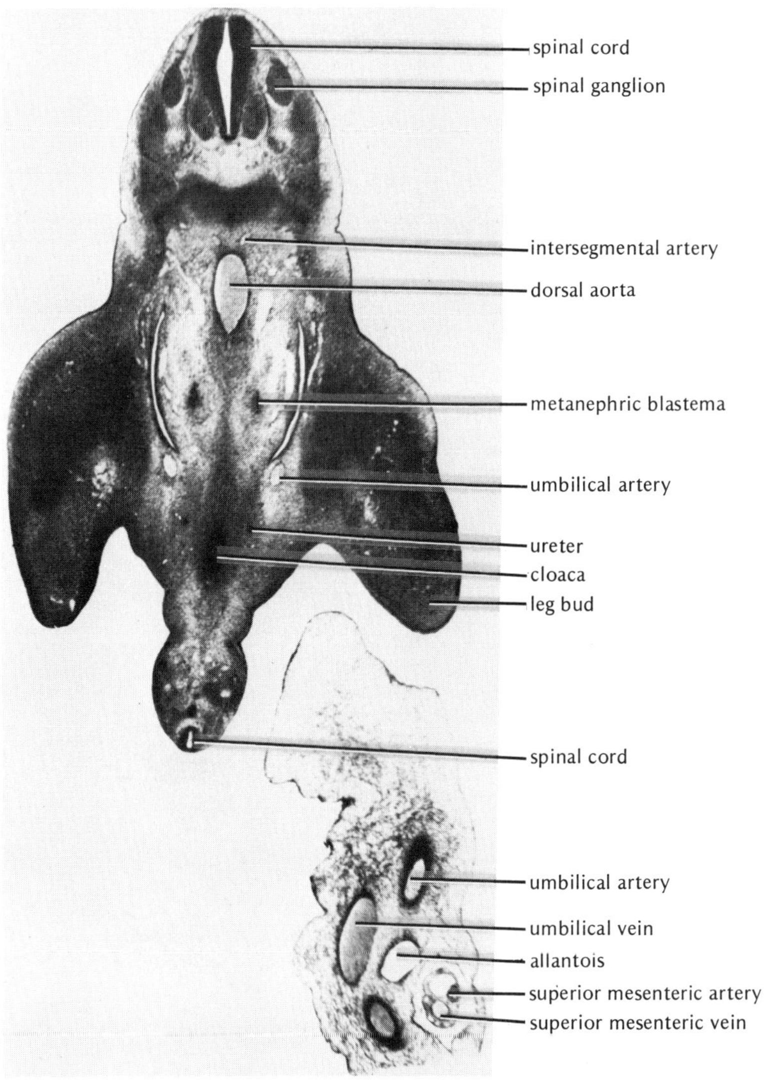

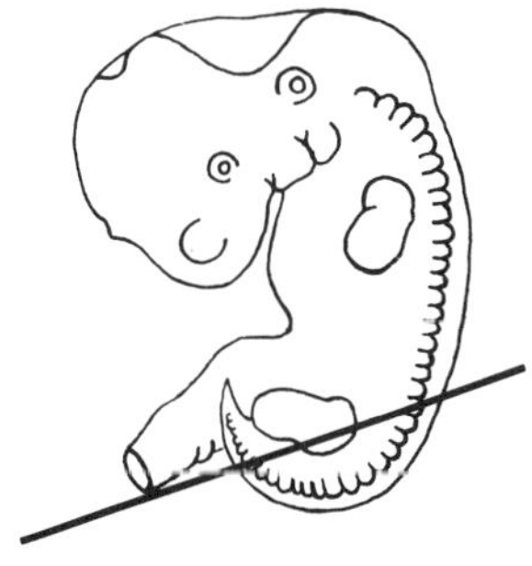

Figure 34 11.5-mm human embryo, transverse section at leg level

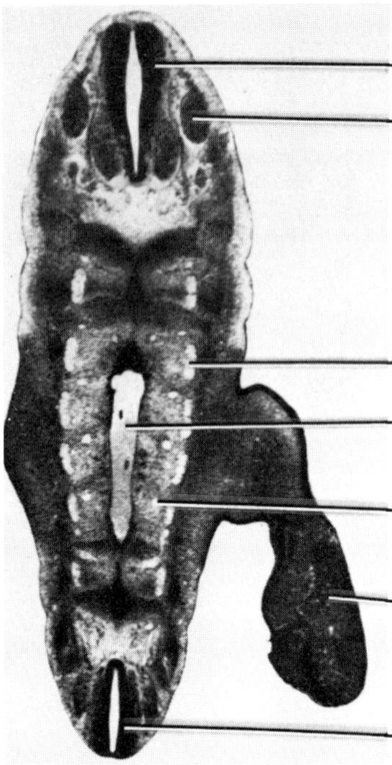

Figure 35 11.5-mm human embryo, transverse section at inferior dorsal aortal level

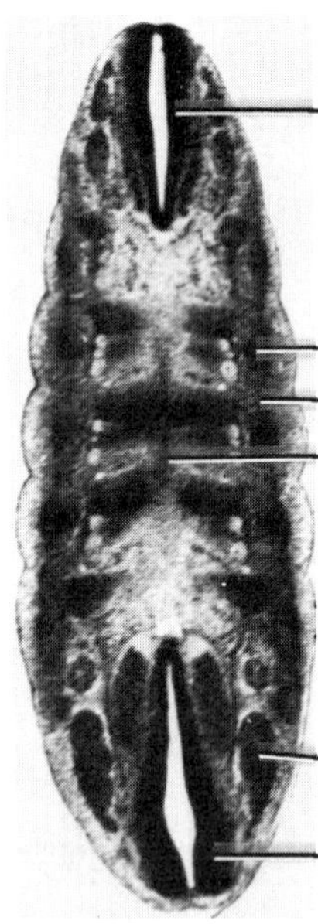

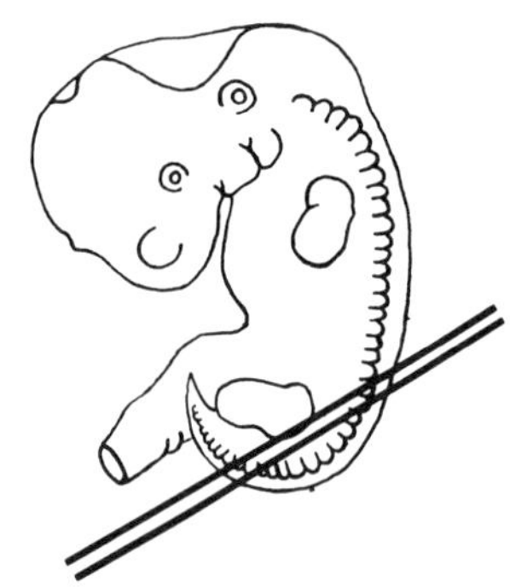

Figure 36 11.5-mm human embryo, transverse section at inferior neural tube level

18.0-mm Human Embryo

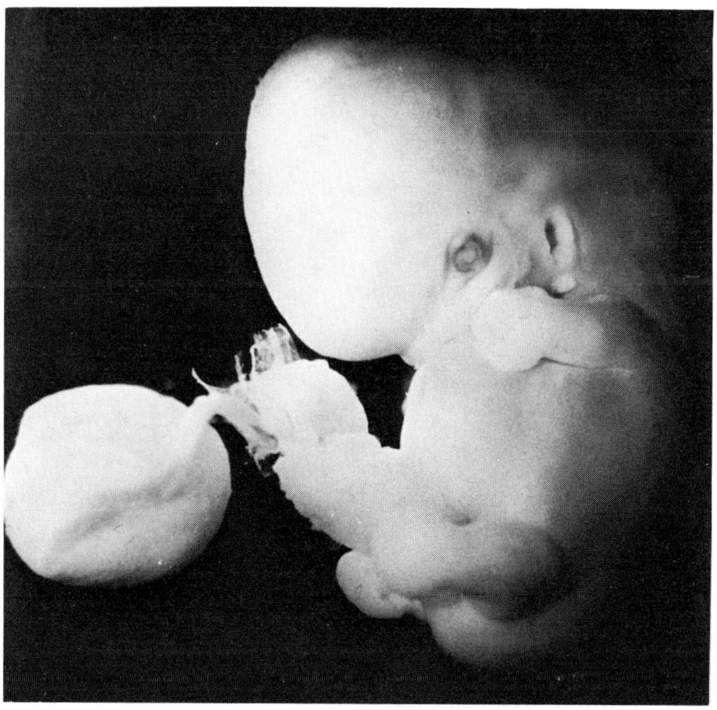

Figure 37 18-mm human embryo

18.0-mm Human Embryo

The specimen chosen to represent this stage is H-203, a female. The embryo was preserved in Lillies fixative (with some resultant shrinkage), sectioned at 10μ, and stained with PAS. The 18-mm human embryo is approximately 42 days (6 weeks) old and is at developmental stage 19.

Following is a list of some of the changes which have occurred since stage 17 (11.5 mm). The trunk is becoming more elongated and straighter. The limbs extend straight forward, the hand rays are notched, and the foot rays are present. The eyelids are appearing and the external ear is reaching its definitive shape. The muscles have begun to develop. The corpus striatum and thalamus are apparent in the brain. Valves have developed in the heart. The duodenum has become occluded. Tracheal cartilage has appeared. Metanephric tubules are developing and the cloacal septum has developed.

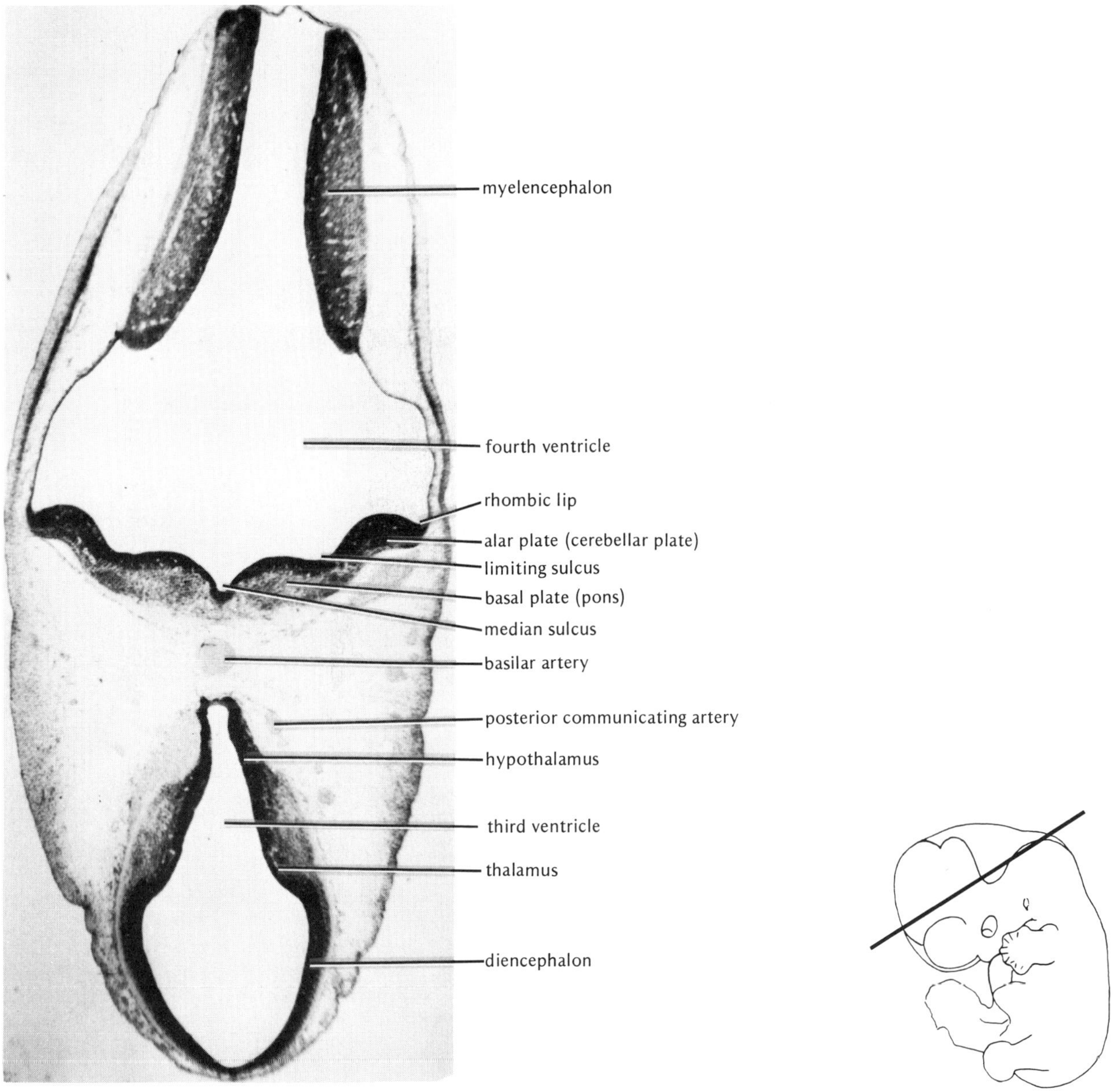

Figure 38 18-mm human embryo, transverse section at brain level

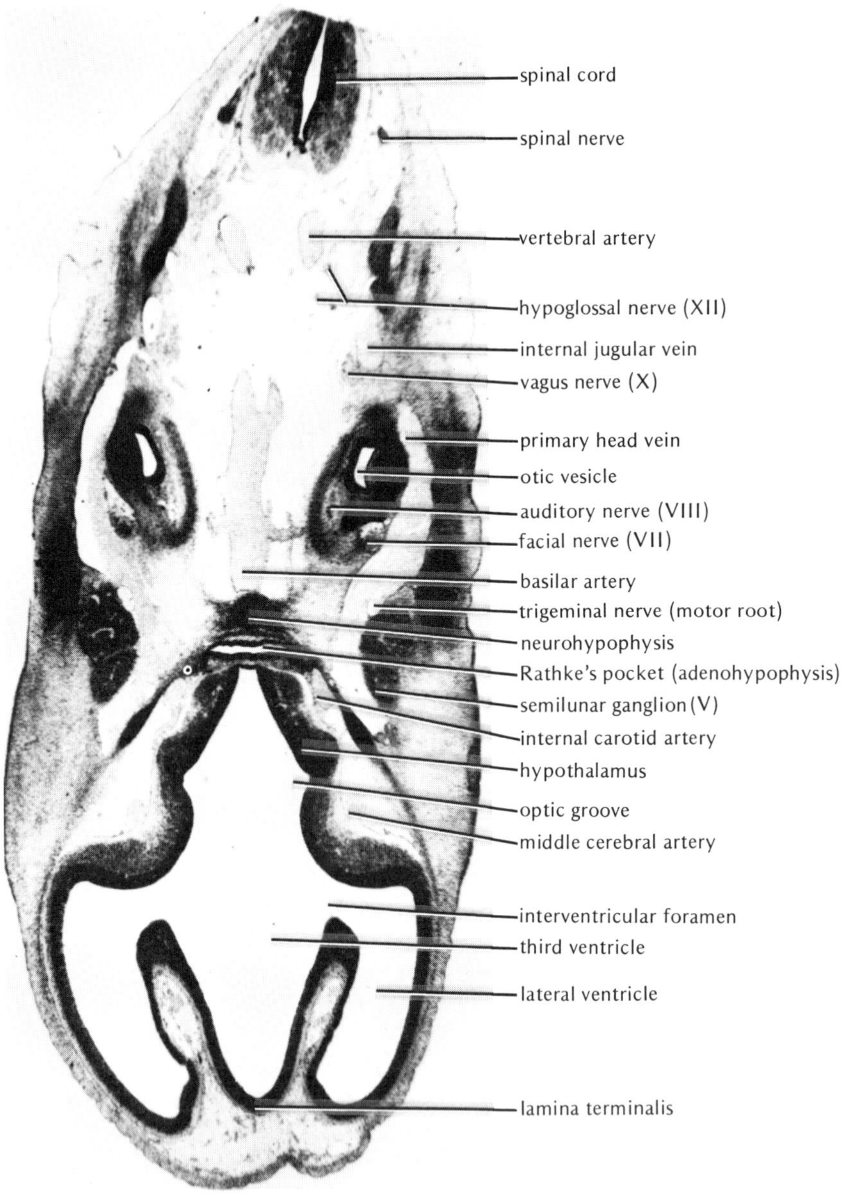

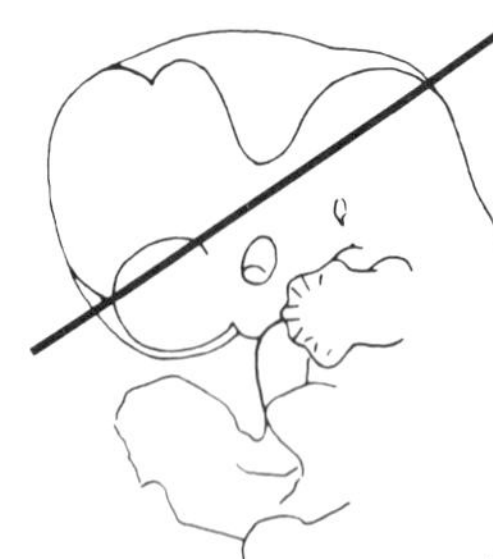

Figure 39 18-mm human embryo, transverse section at interventricular foramen level

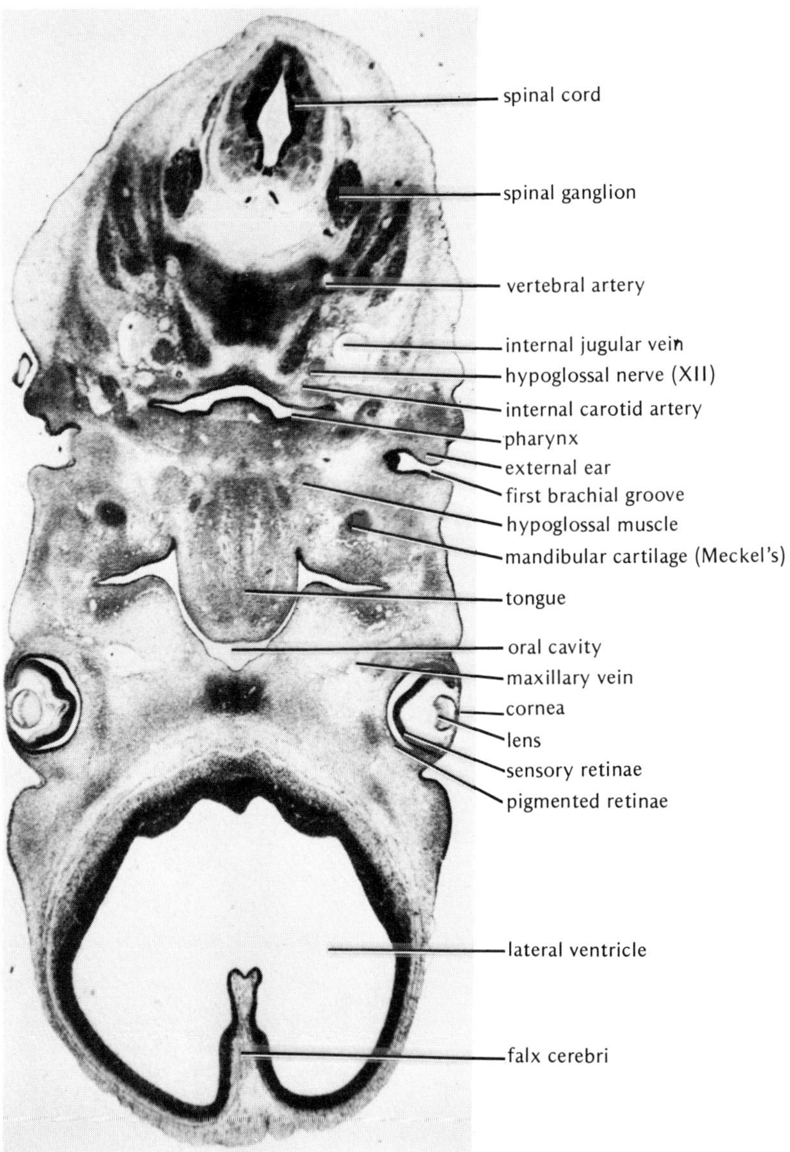

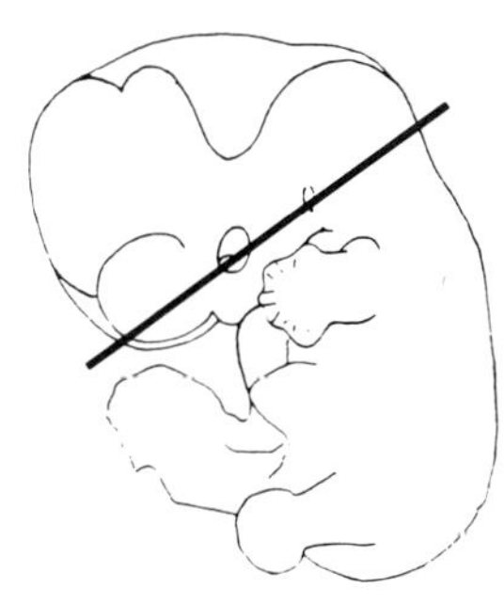

Figure 40 18-mm human embryo, transverse section at eye level

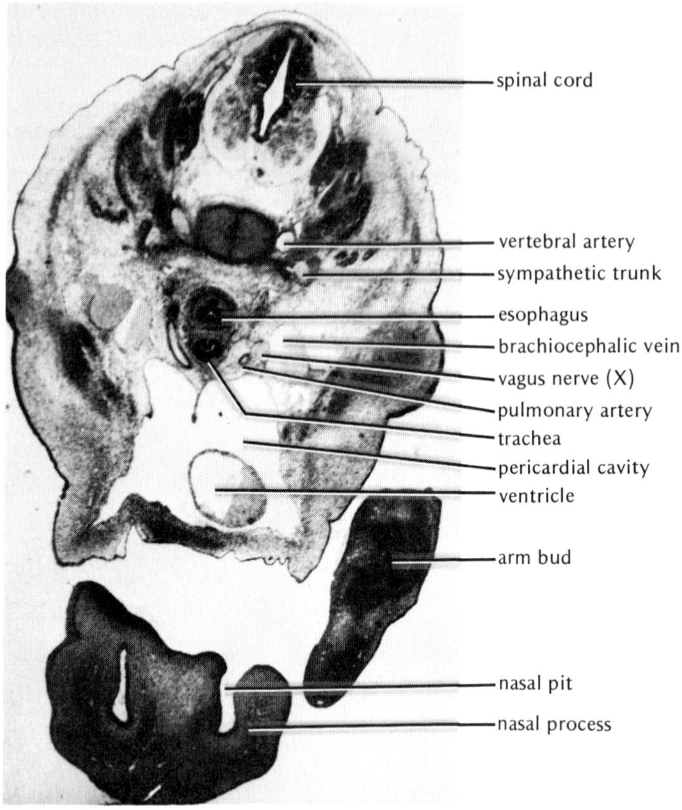

Figure 41 18-mm human embryo, transverse section at external nares level

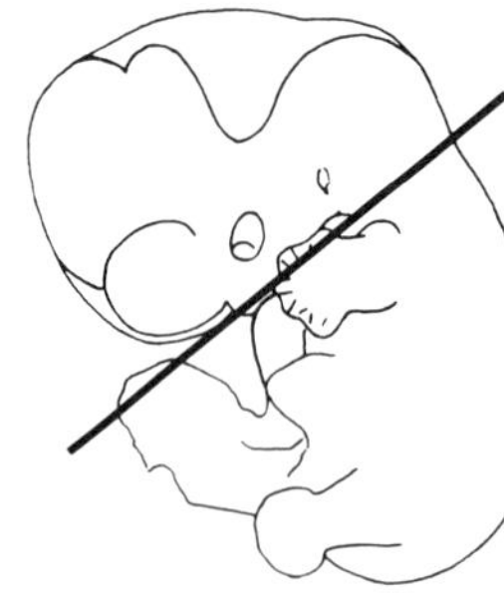

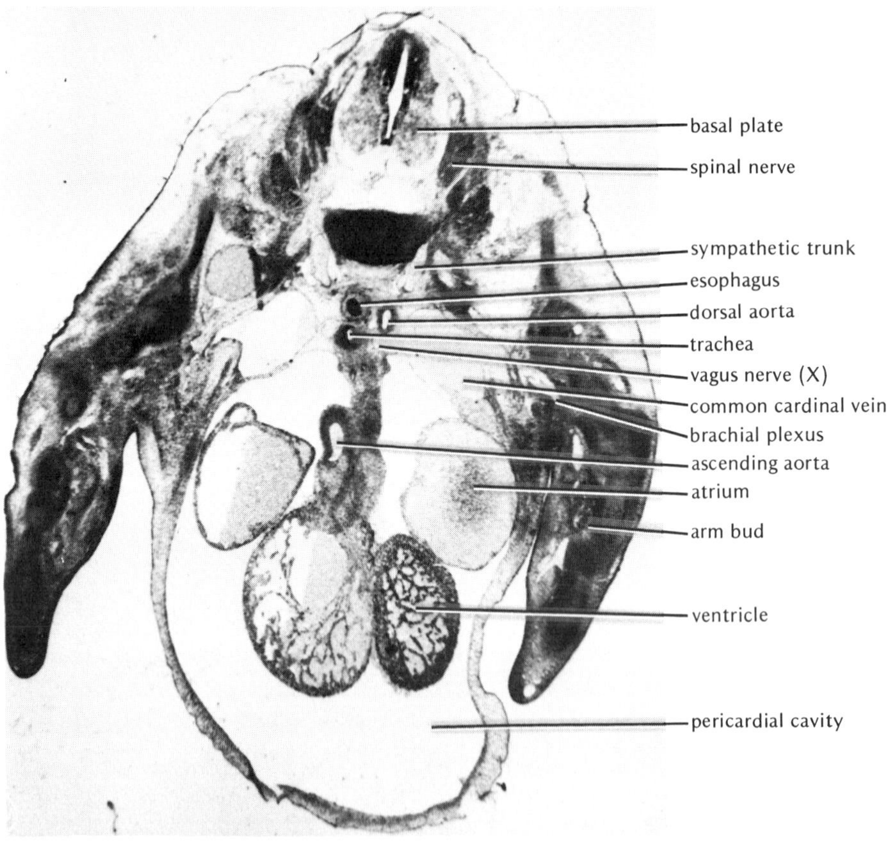

Figure 42 18-mm human embryo, transverse section at arm level

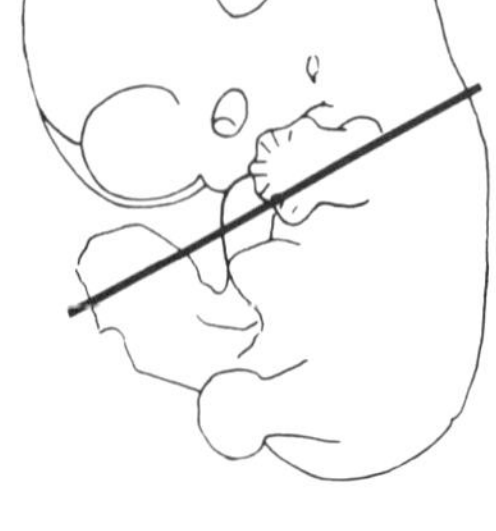

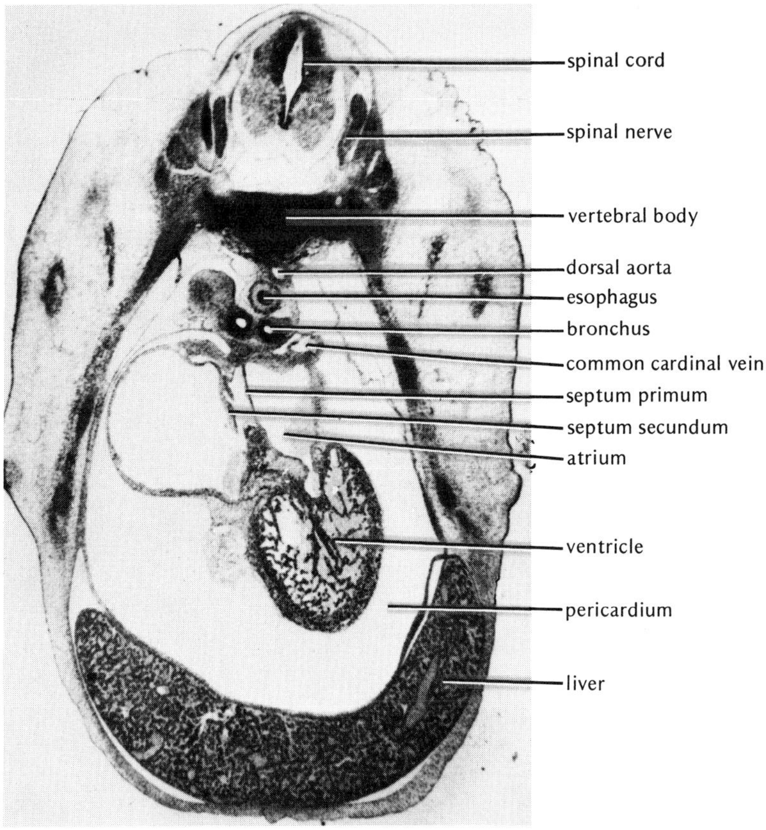

Figure 43 18-mm human embryo, transverse section at heart level

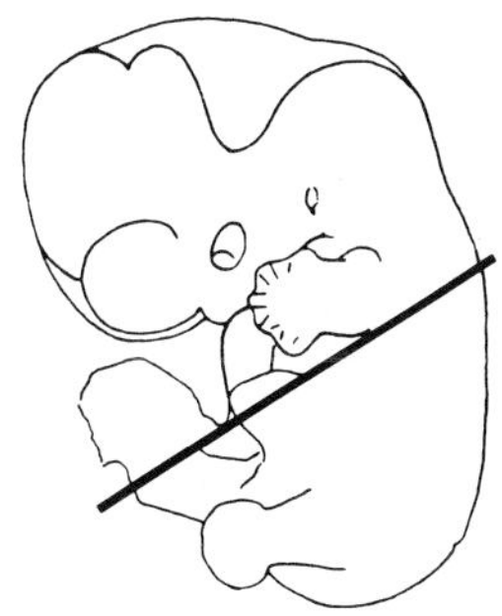

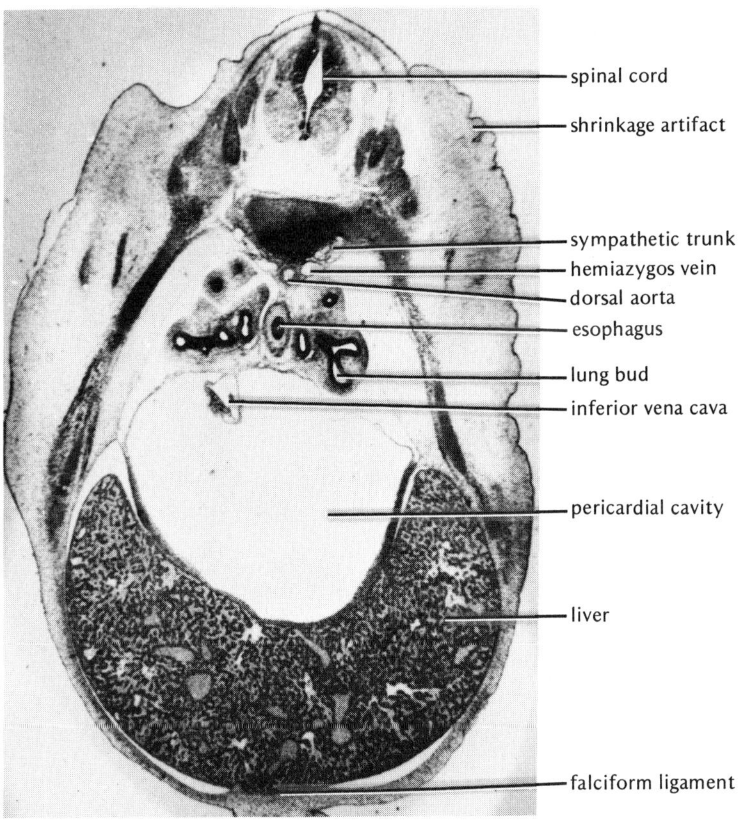

Figure 44 18-mm human embryo, transverse section at liver level

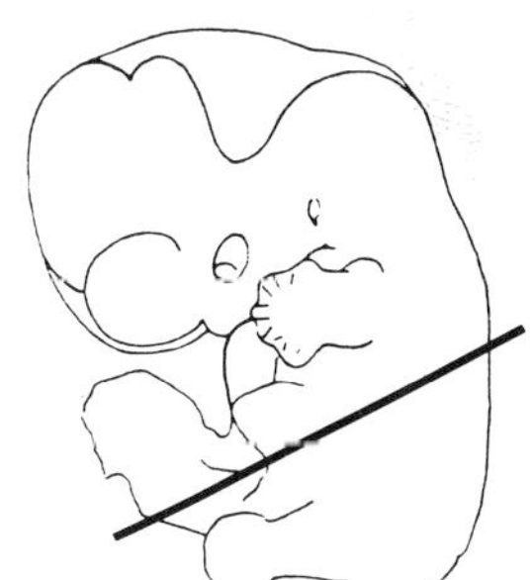

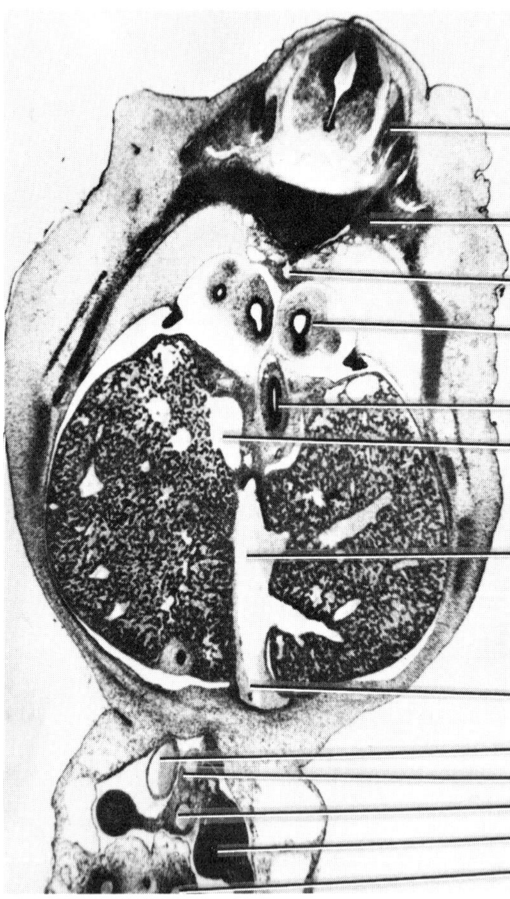

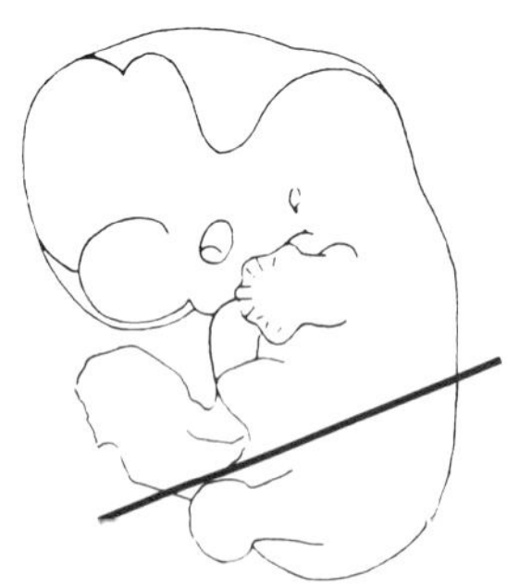

Figure 45 18-mm human embryo, transverse section at umbilical hernia level

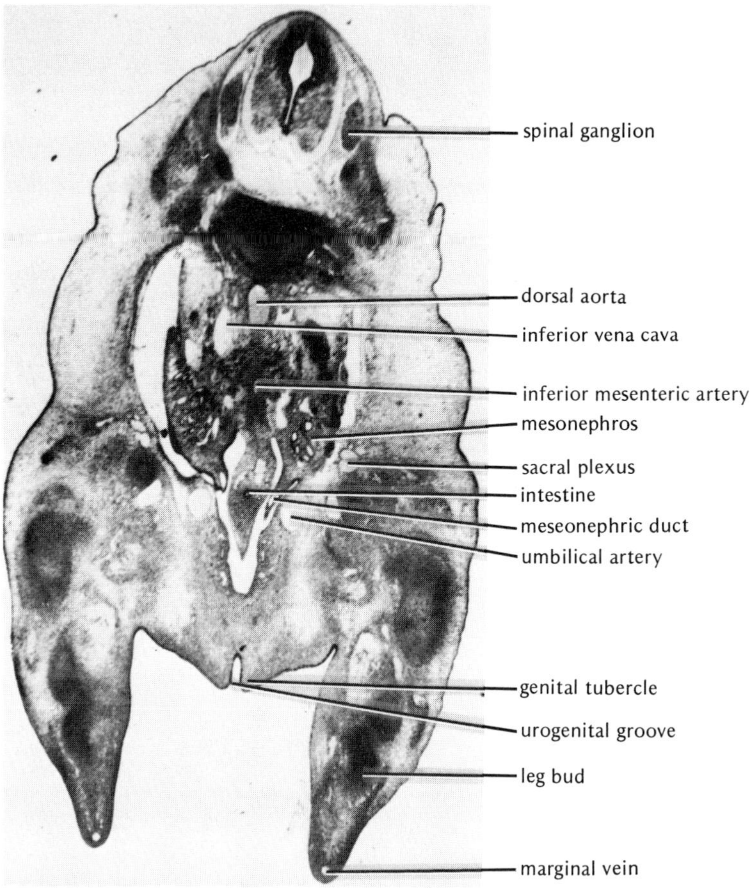

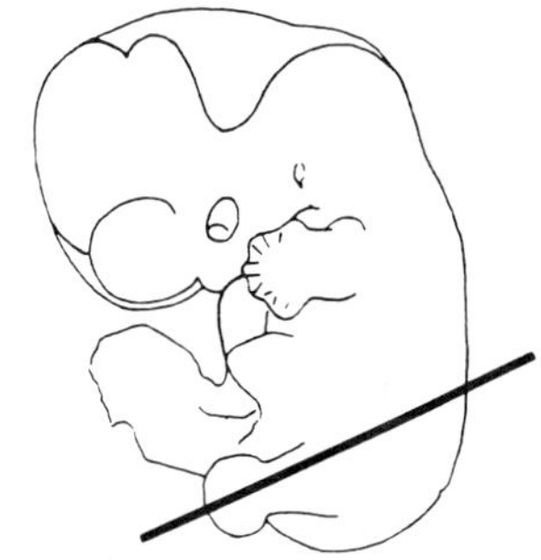

Figure 46 18-mm human embryo, transverse section at leg level

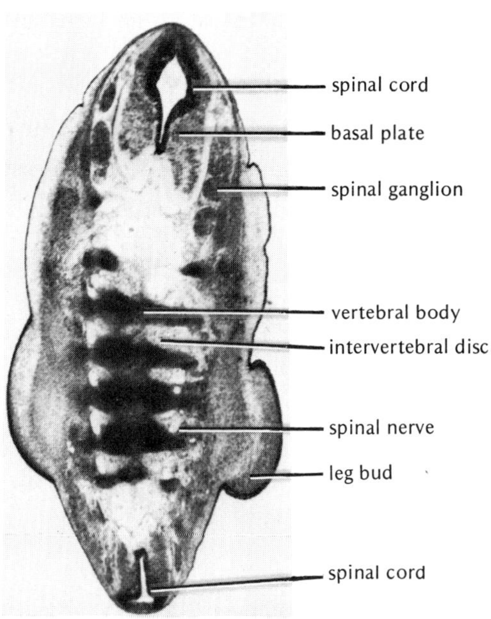

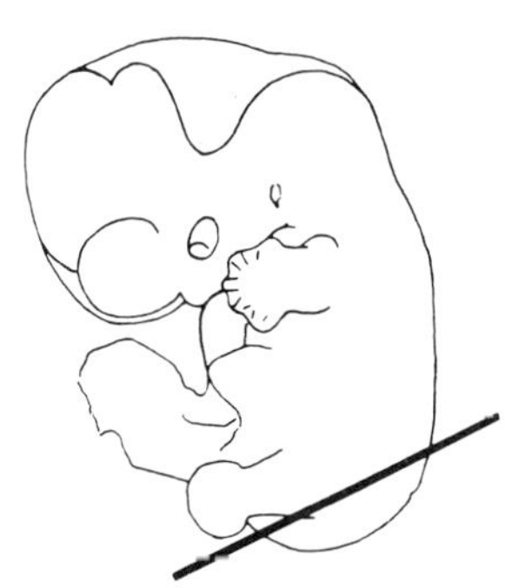

Figure 47 18-mm human embryo, transverse section at tail bud level

32.0-mm Human Embryo

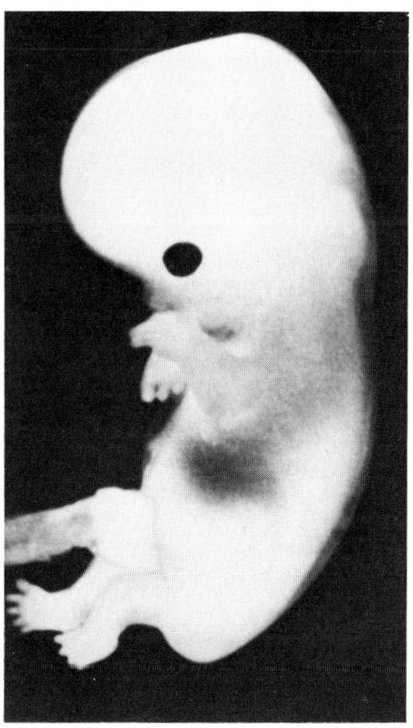

Figure 48 32-mm human embryo

32.0-mm Human Embryo

The specimen chosen to represent this stage is H-727, a female, weighing 3.5 gm, and having a foot length of 4.5 mm. The embryo was preserved in Lillies fixative, sectioned at 10μ, and stained with PAS.

This embryo represents the end of the embryonic period. The 32 mm human embryo is approximately 54 days (8 weeks) old and is slightly more advanced than the oldest horizon described by Streeter (which was stage 23). At the end of the embryonic period and beginning of the fetal period (usually set at 60 days) all organ systems are present and are represented approximately in their definitive form.

Following is a list of some of the changes which have occurred since stage 19 (18 mm). The head is more erect and rounded. The limbs and digits have elongated to the point where the fingers overlap those of the other hand and the toes touch those of the other foot. Bone formation has begun in the long bones, muscles are well developed, and movement begins. The eyelids have developed, the cornea has thinned out, and the optic nerve has attained its definitive form. The cochlea exhibits two complete spiral turns. The adenohypophysis has lost its connection to Rathke's pocket. The heart has attained its fetal condition with the addition of the septum secundum. The lungs have attained their definitive shape. The duodenum has reopened. The metanephric kidney is reaching a functional state. The gonads can be distinguished as ovarian and the Müllerian ducts are fusing to form the uterus.

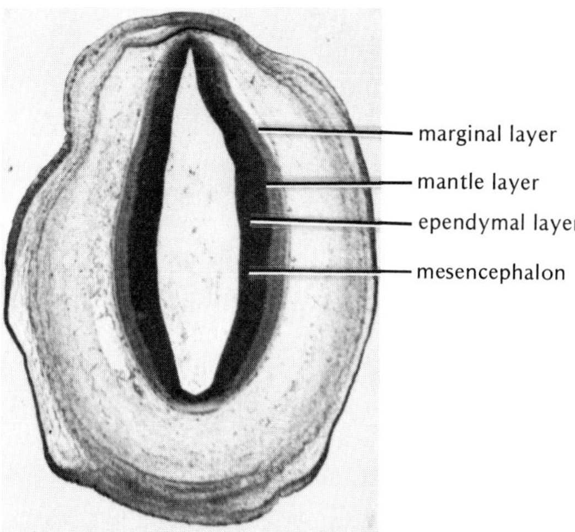

Figure 49 32-mm human embryo, transverse section at mesencephalon level

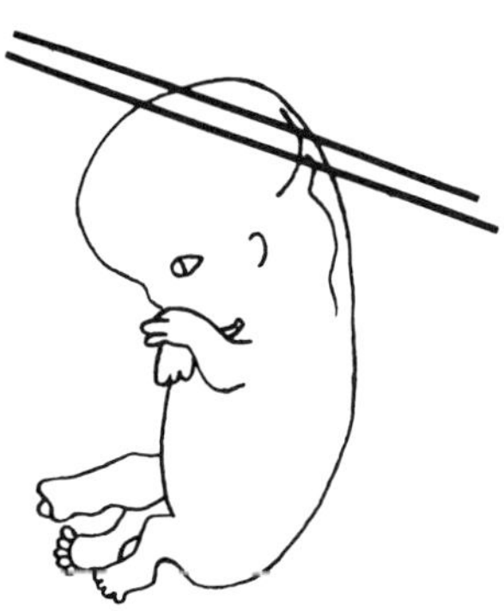

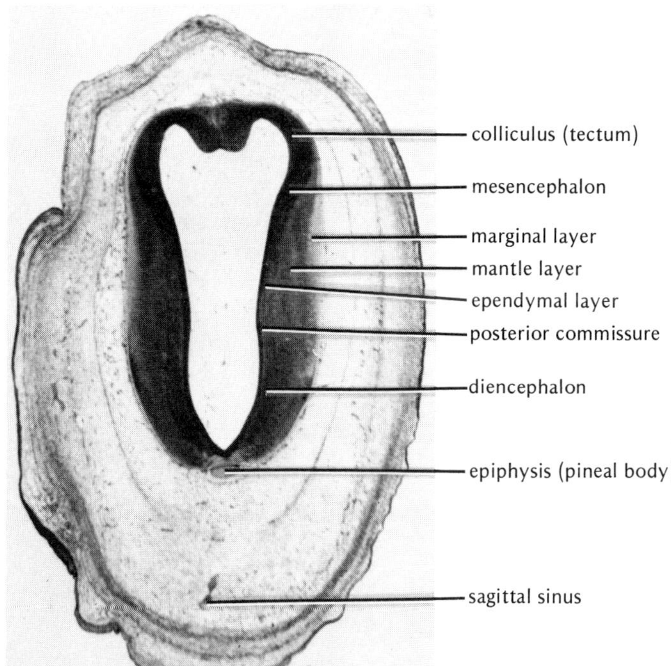

Figure 50 32-mm human embryo, transverse section at tectum level

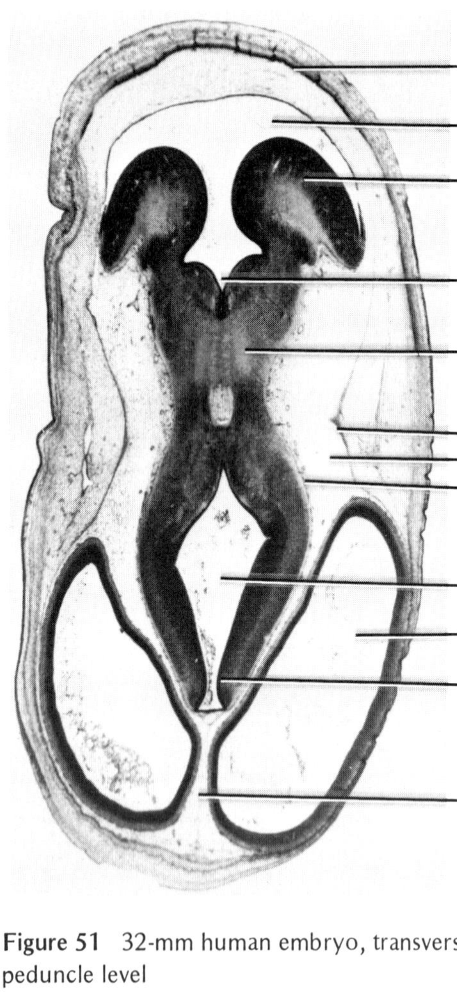

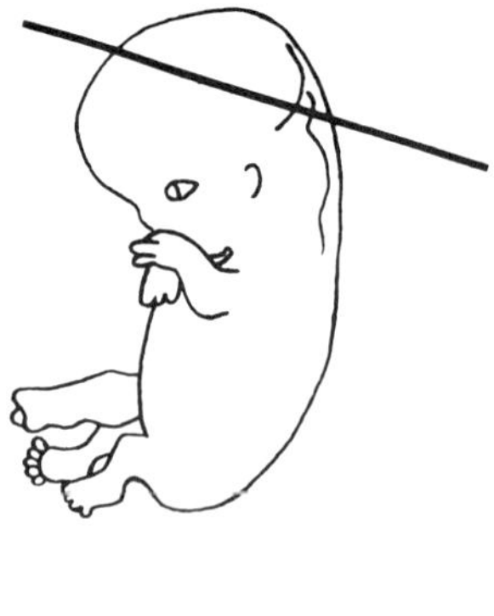

Figure 51 32-mm human embryo, transverse section at peduncle level

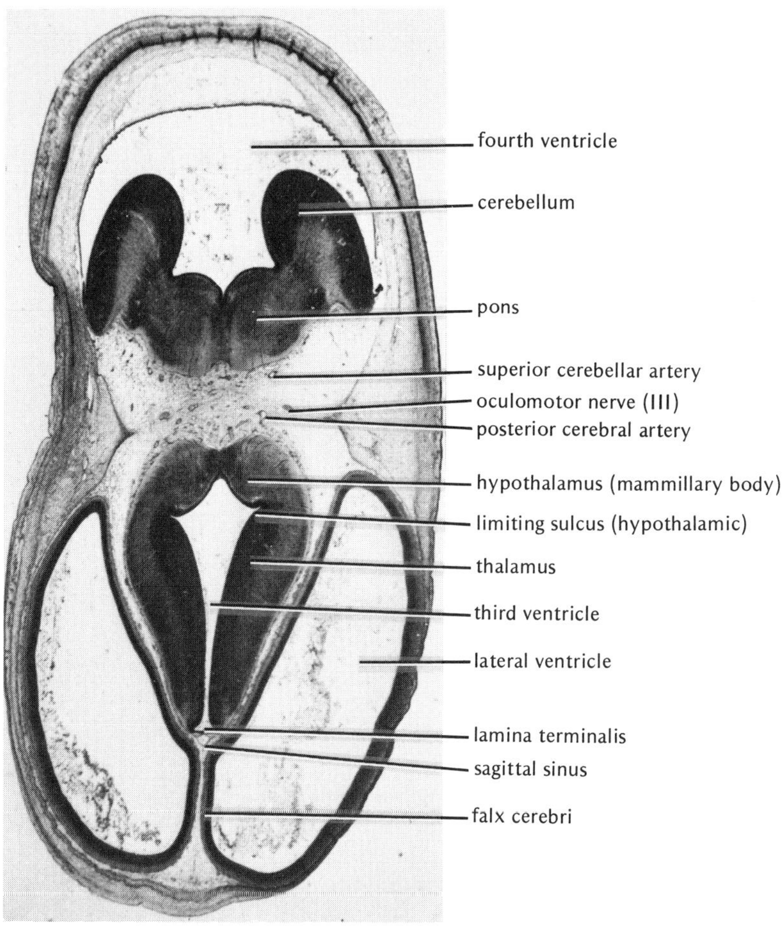

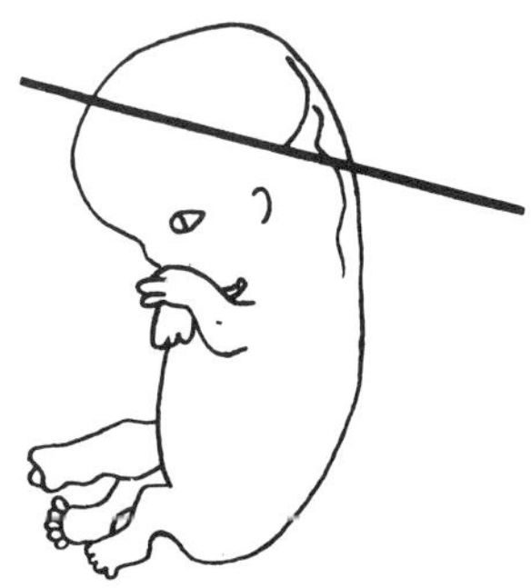

Figure 52 32-mm human embryo, transverse section at cerebellum level

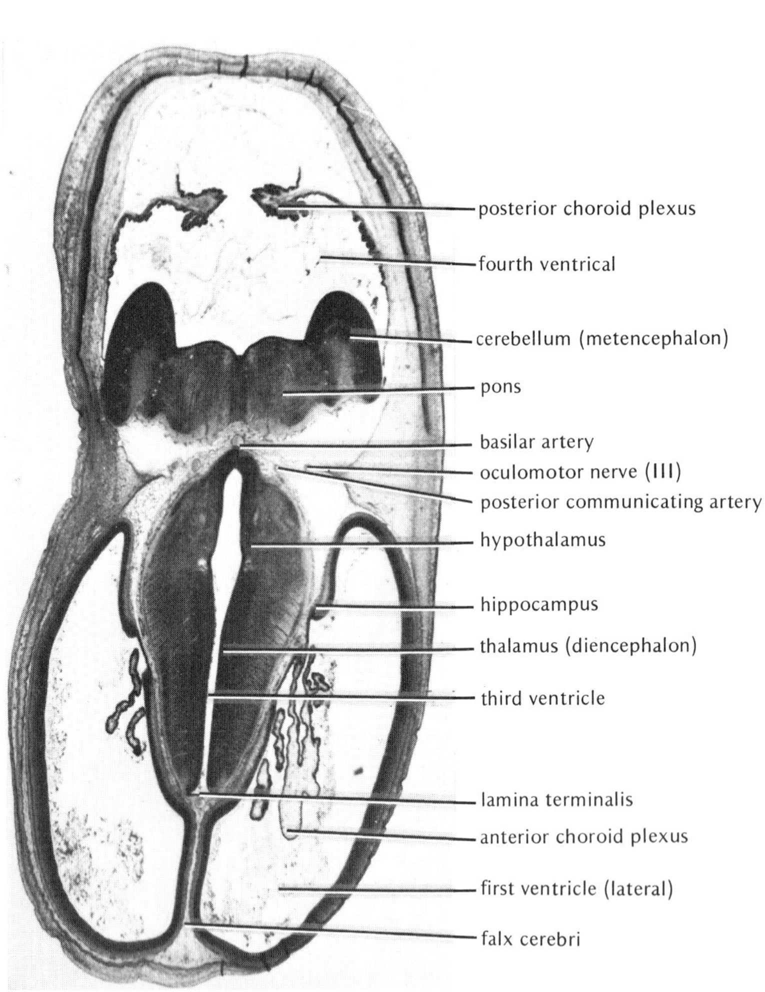

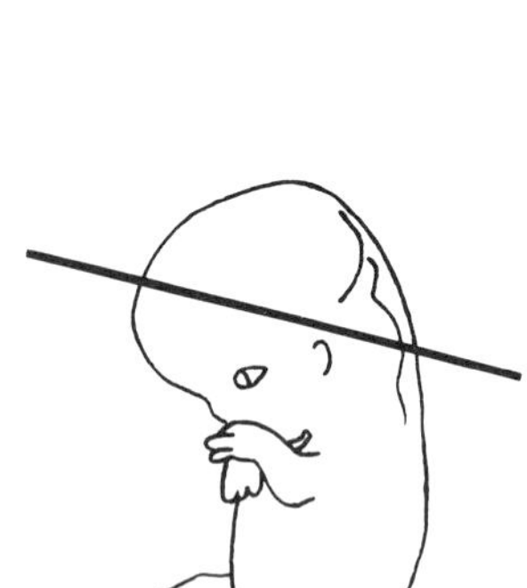

Figure 53 32-mm human embryo, transverse section at pons level

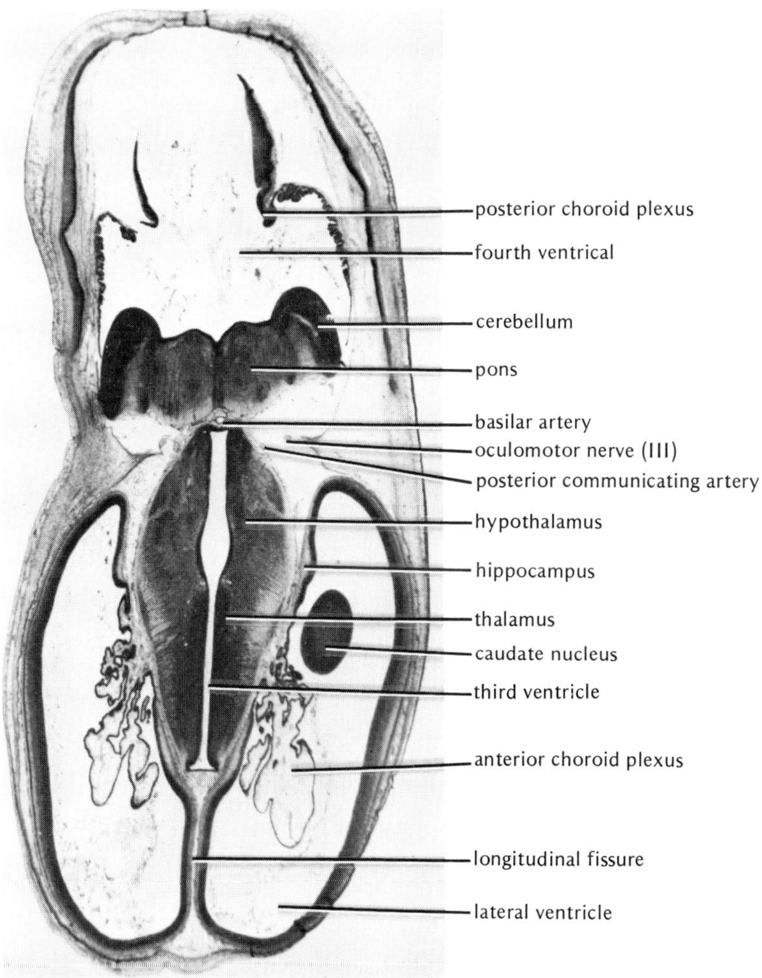

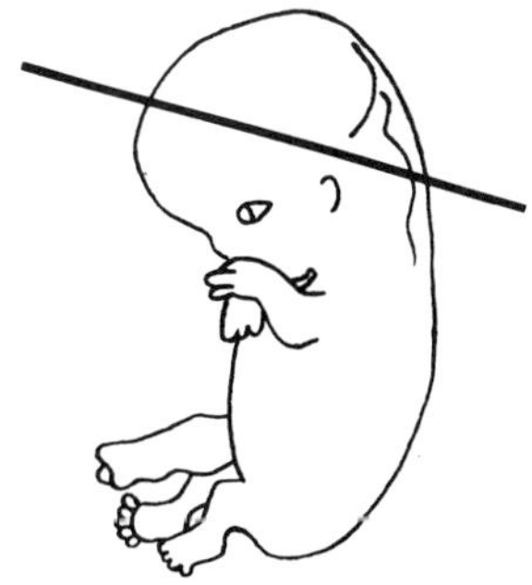

Figure 54 32-mm human embryo, transverse section at hippocampus level

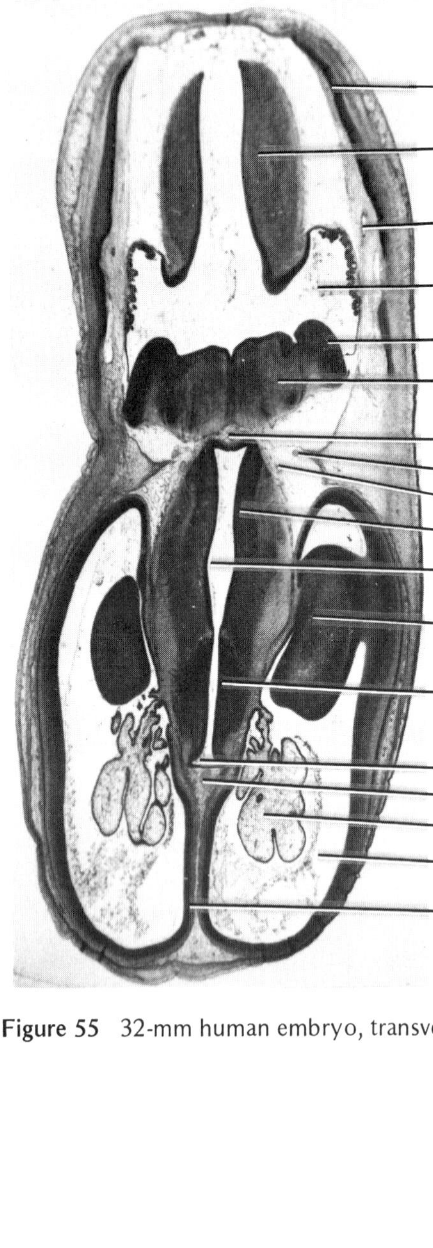

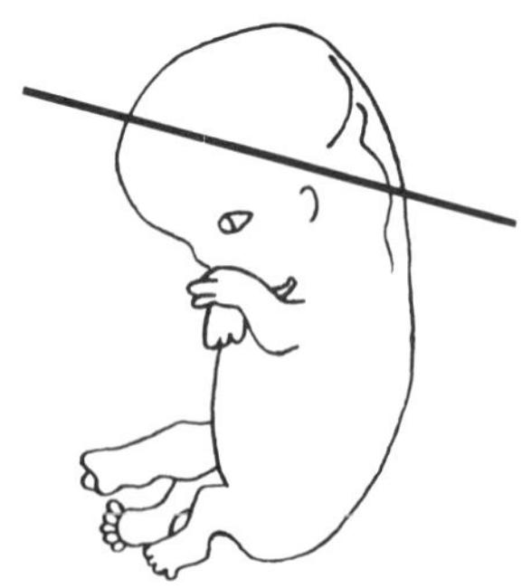

Figure 55 32-mm human embryo, transverse section at choroid plexus

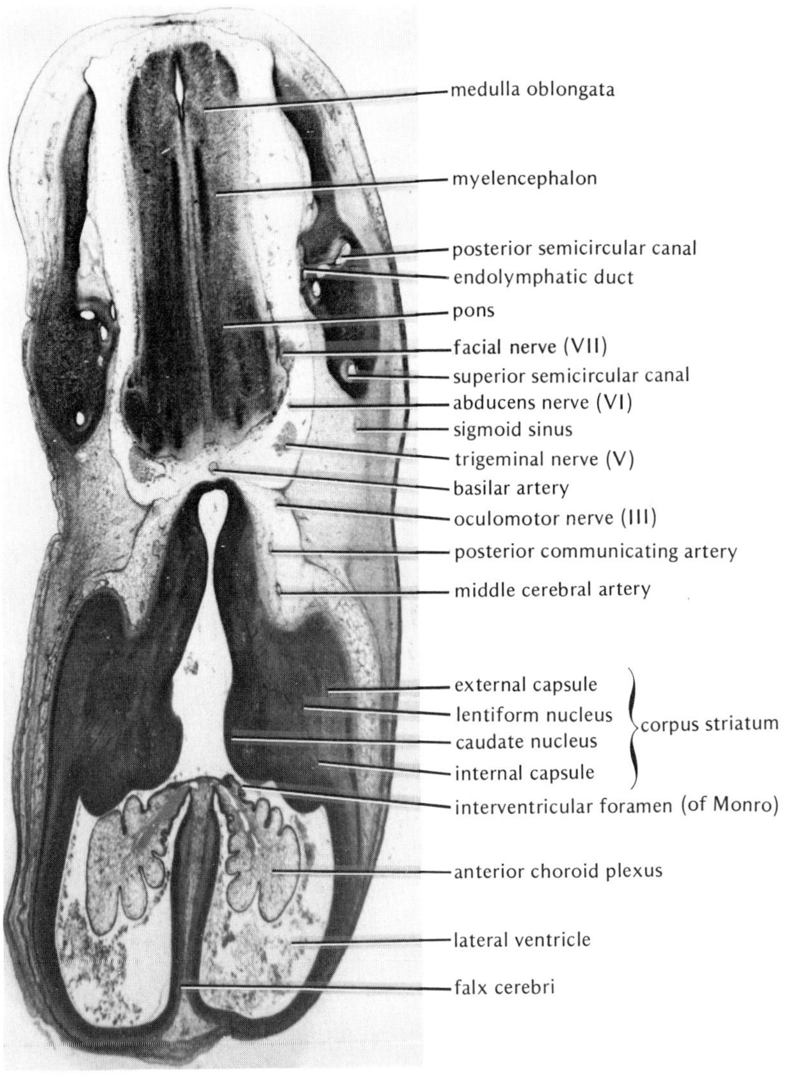

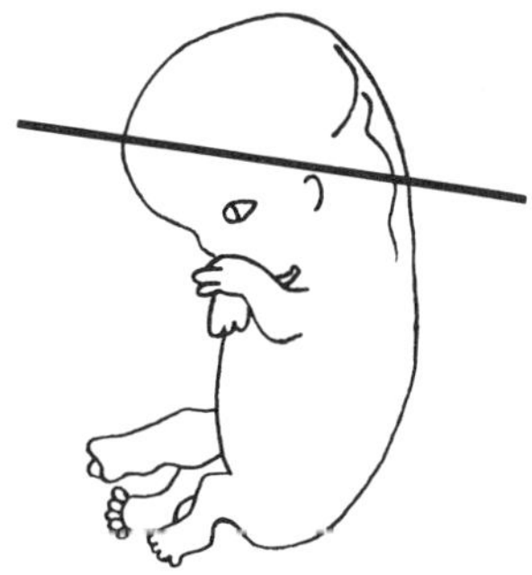

Figure 56 32-mm human embryo, transverse section at corpus striatum level

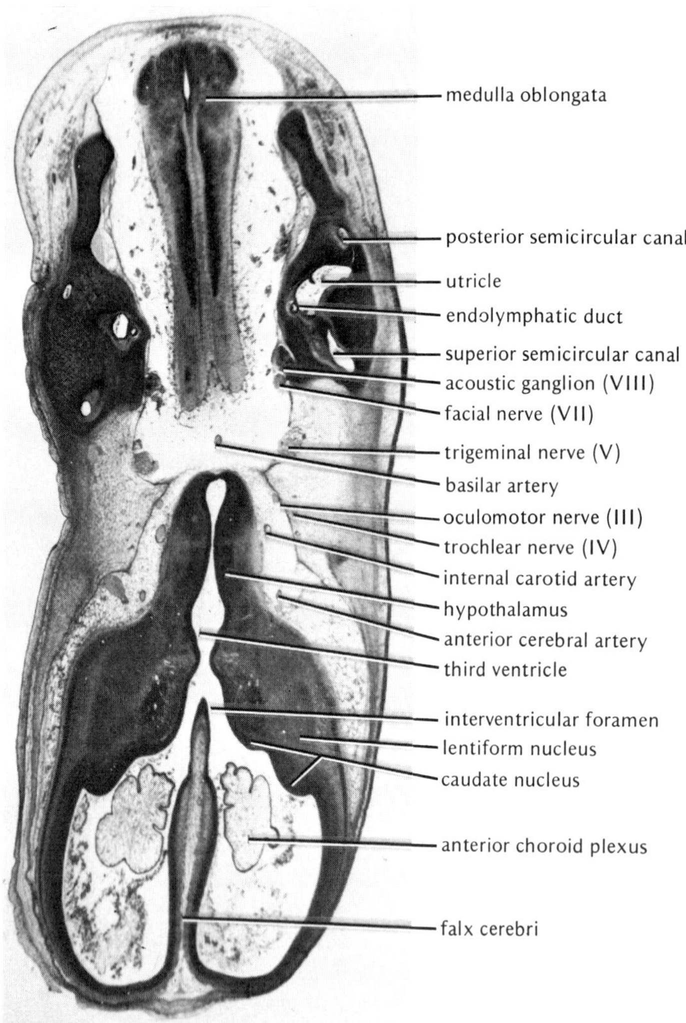

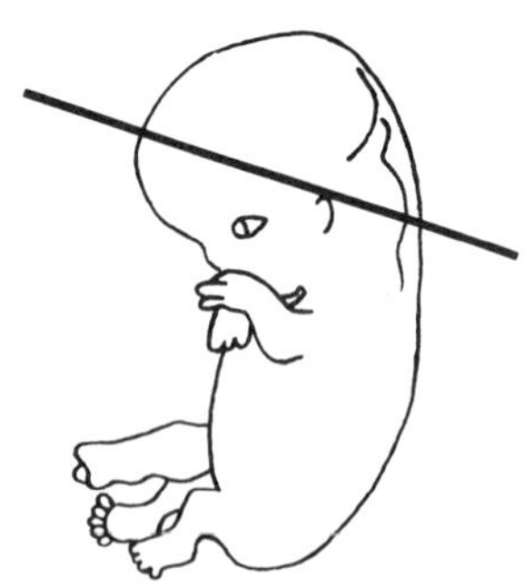

Figure 57 32-mm human embryo, transverse section at interventricular foramen level

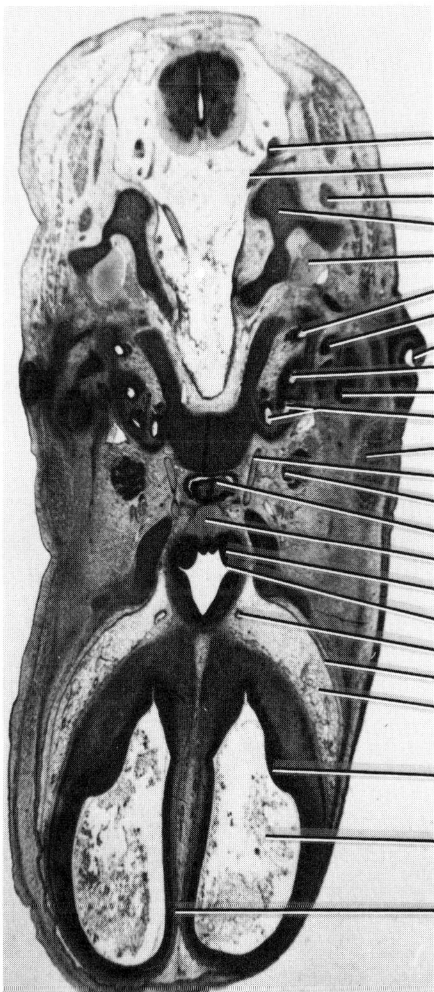

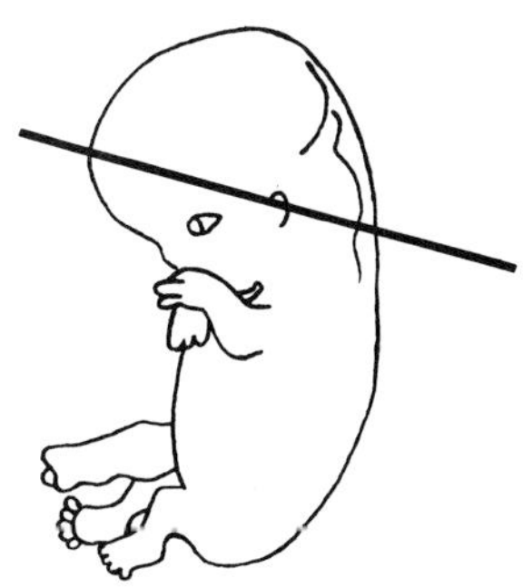

Figure 58 32-mm human embryo, transverse section at inner ear level

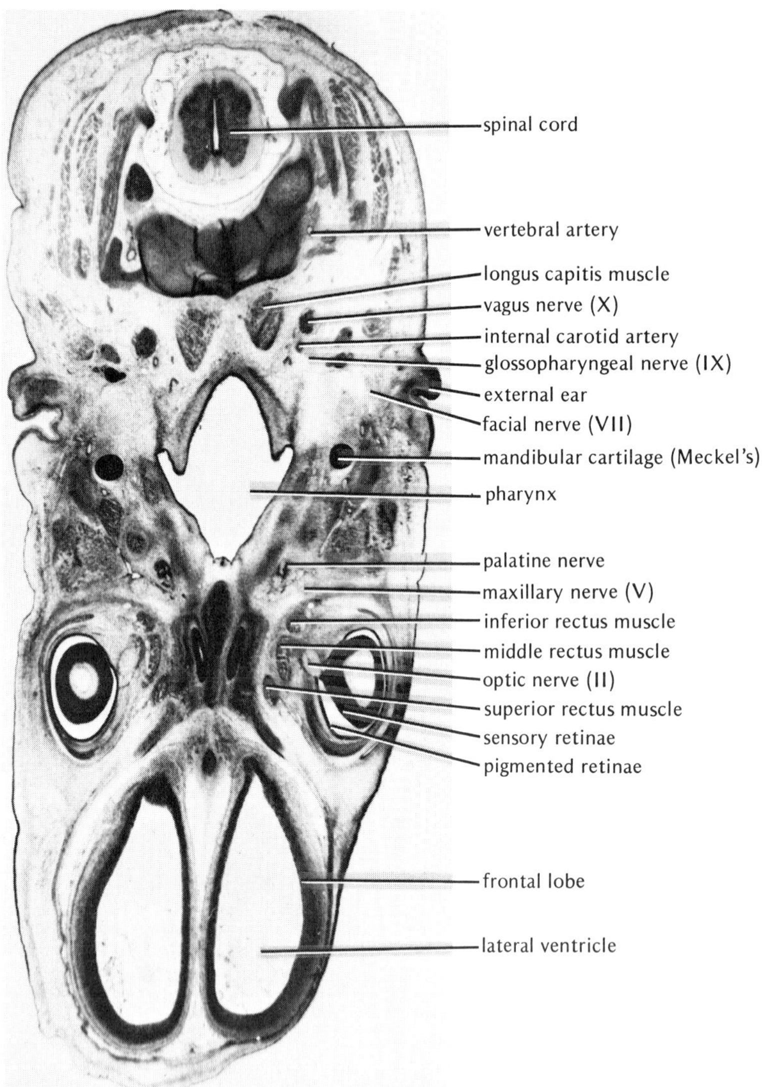

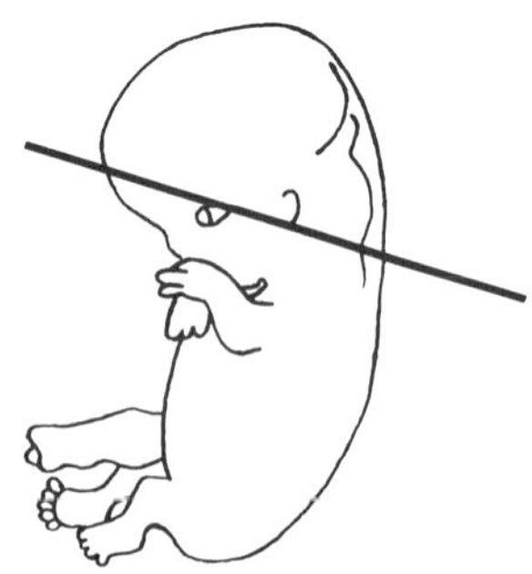

Figure 59 32-mm human embryo, transverse section at pharynx level

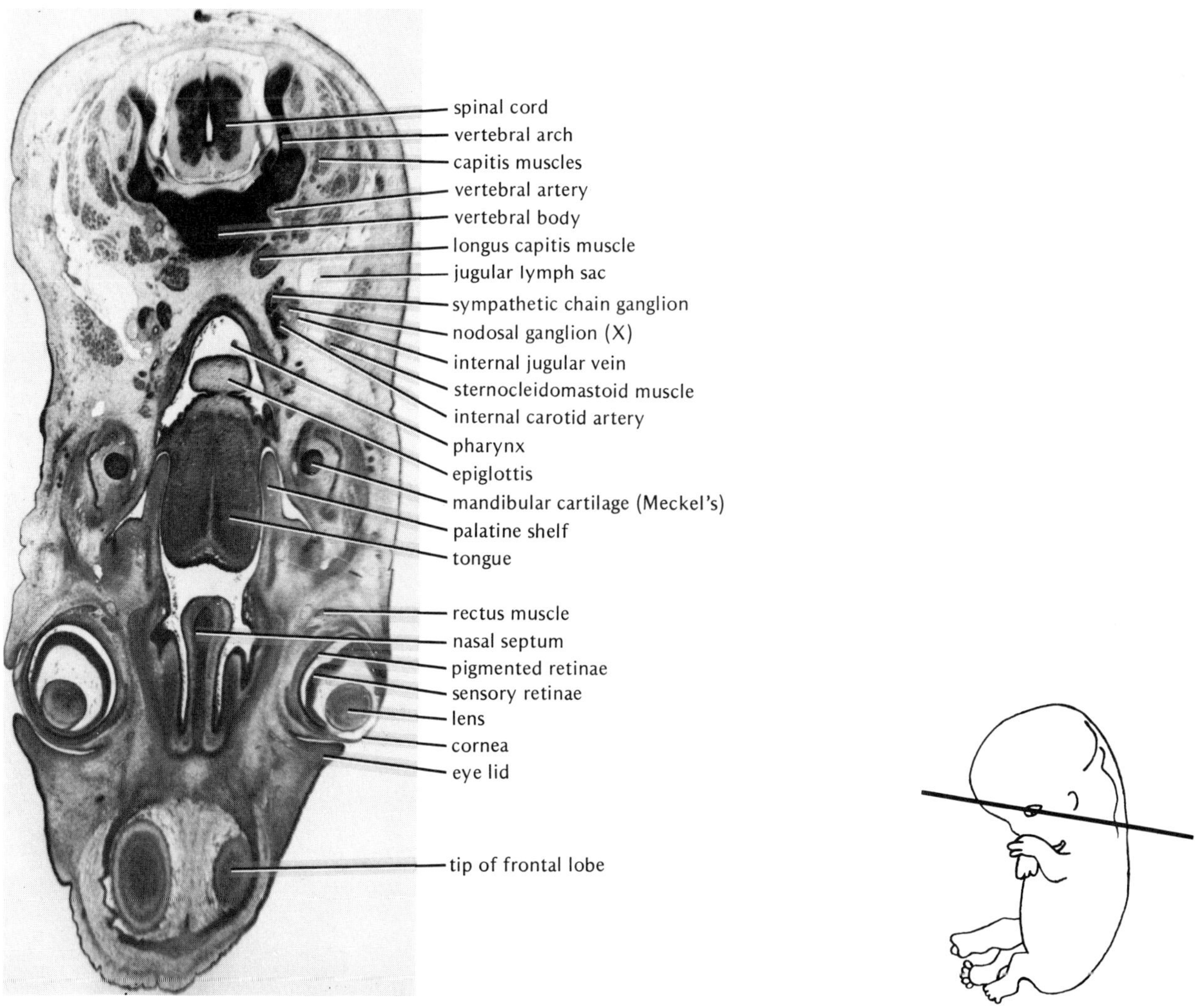

Figure 60 32-mm human embryo, transverse section at eye level

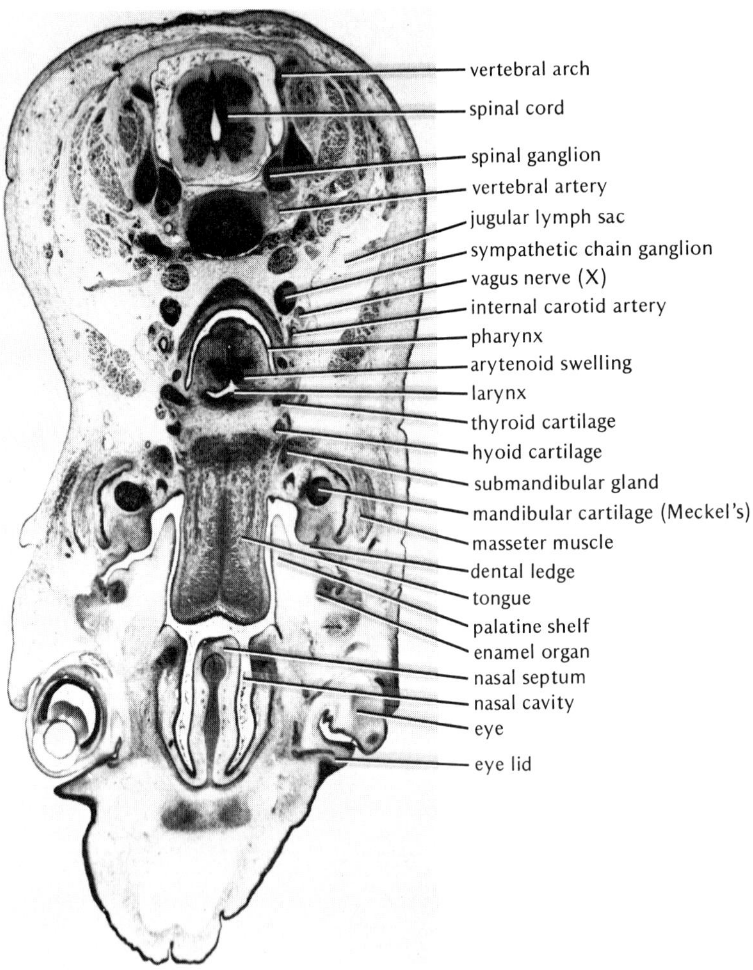

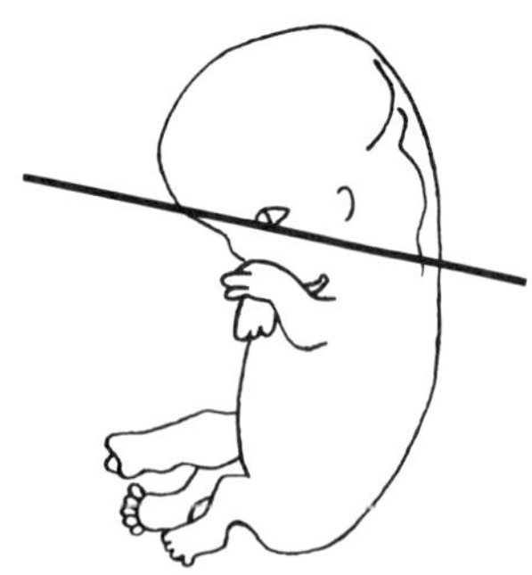

Figure 61 32-mm human embryo, transverse section at palate level

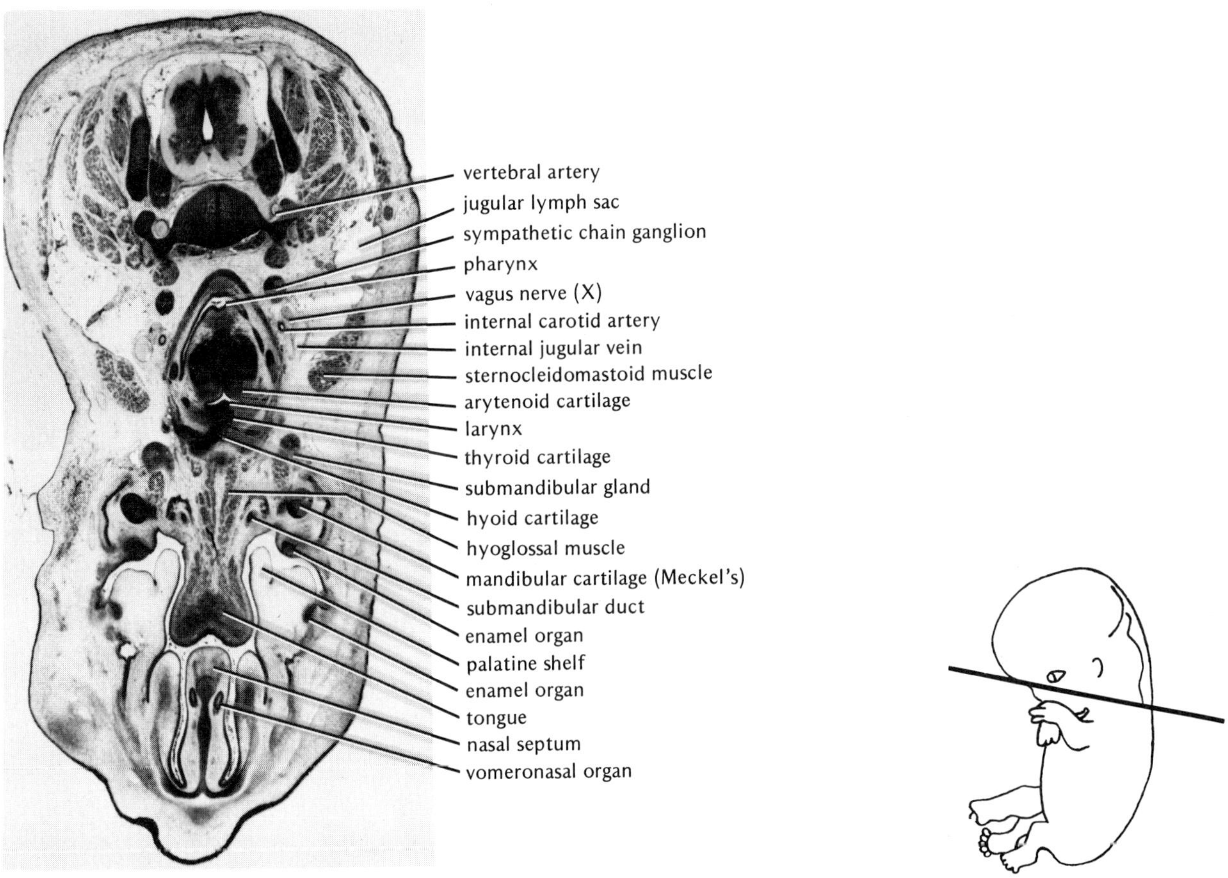

Figure 62 32-mm human embryo, transverse section at nasal level

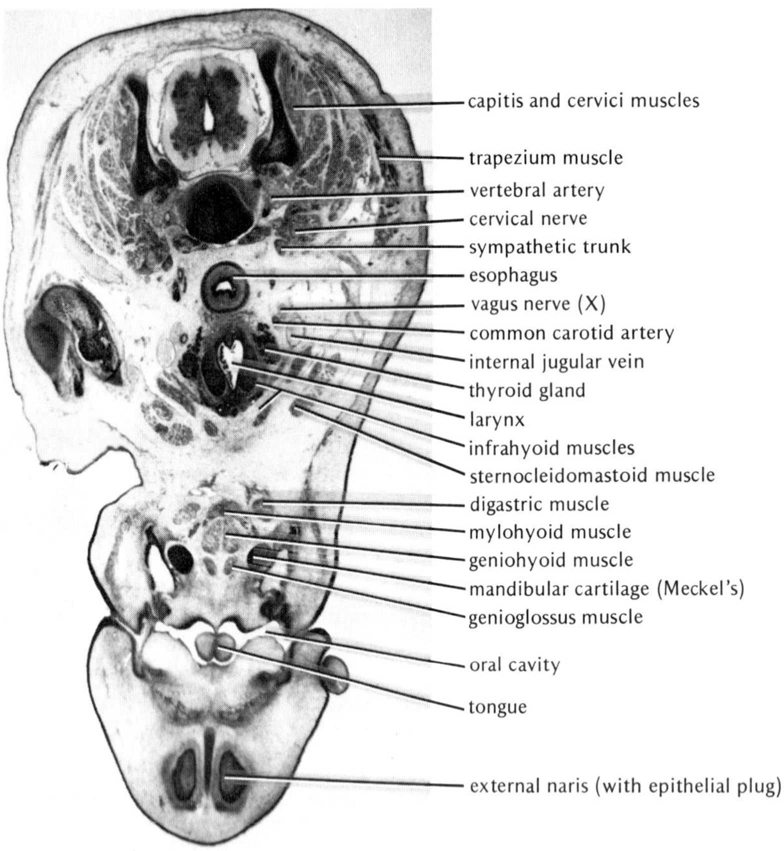

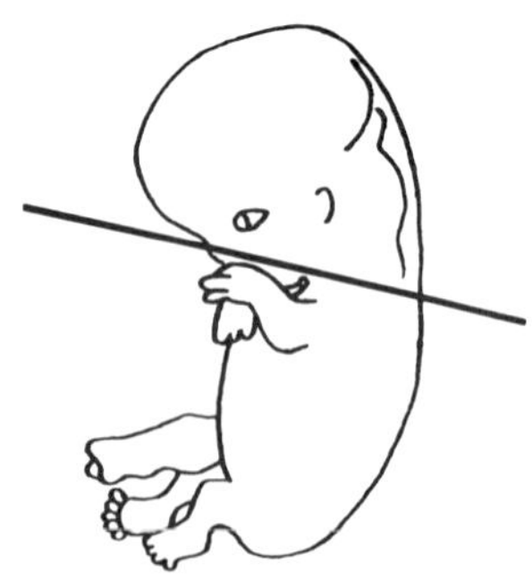

Figure 63 32-mm human embryo, transverse section at maxillary level

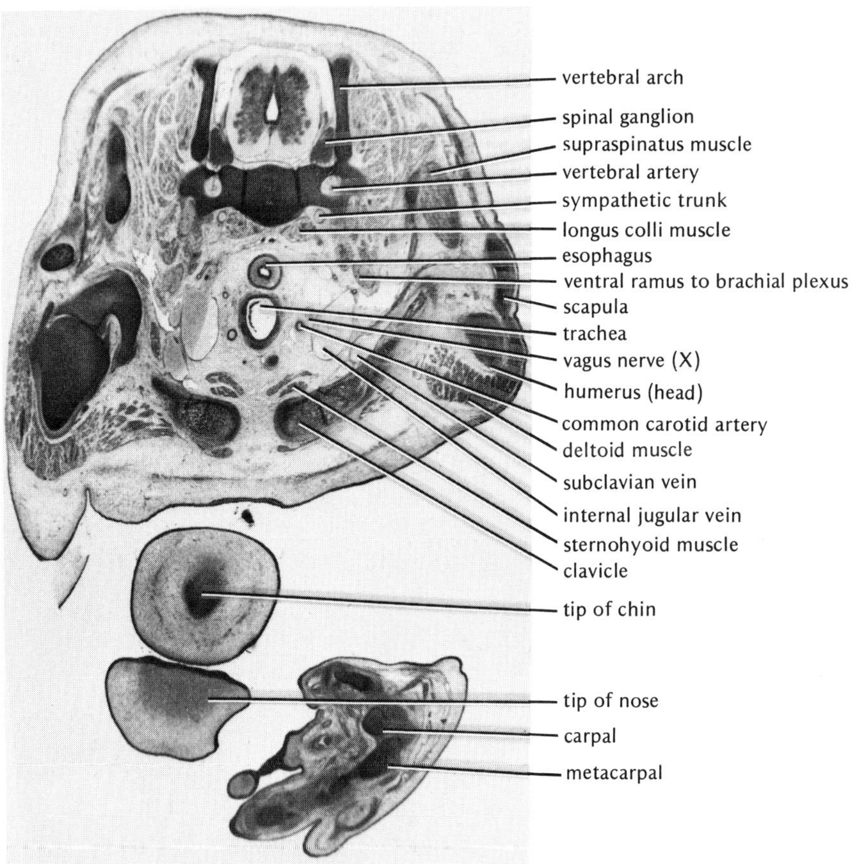

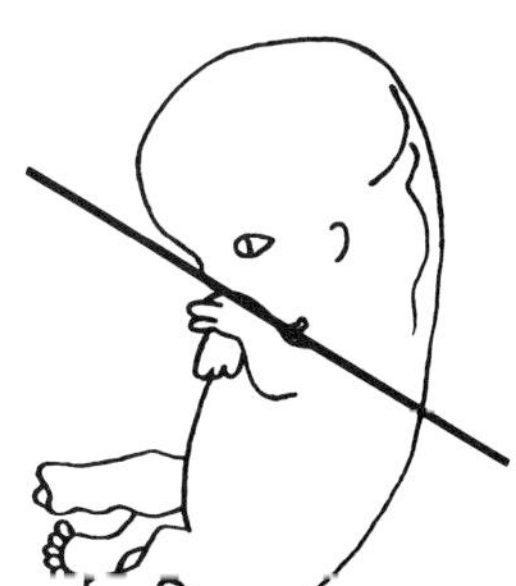

Figure 64 32-mm human embryo, transverse section at shoulder level

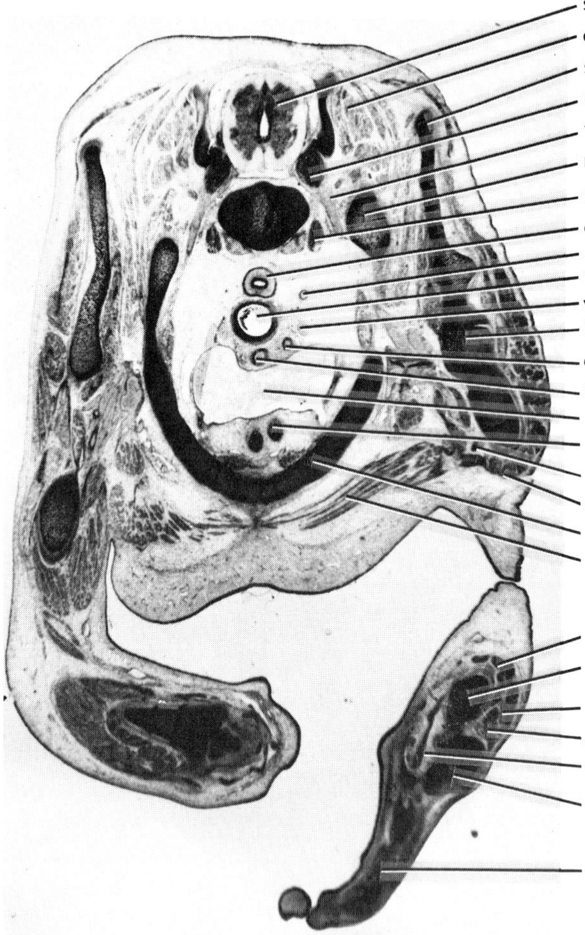

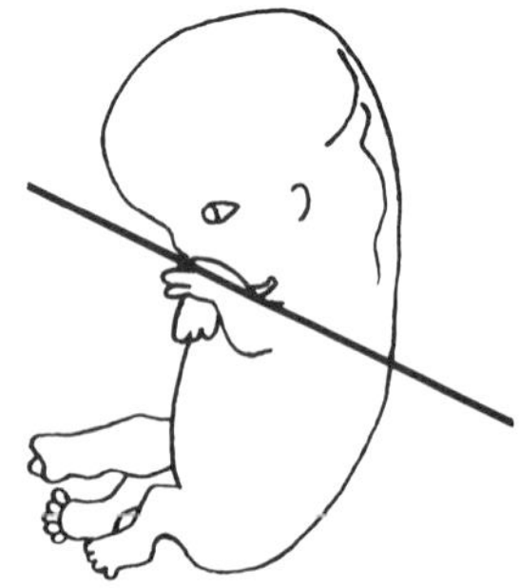

Figure 65 32-mm human embryo, transverse section at upper thorax level

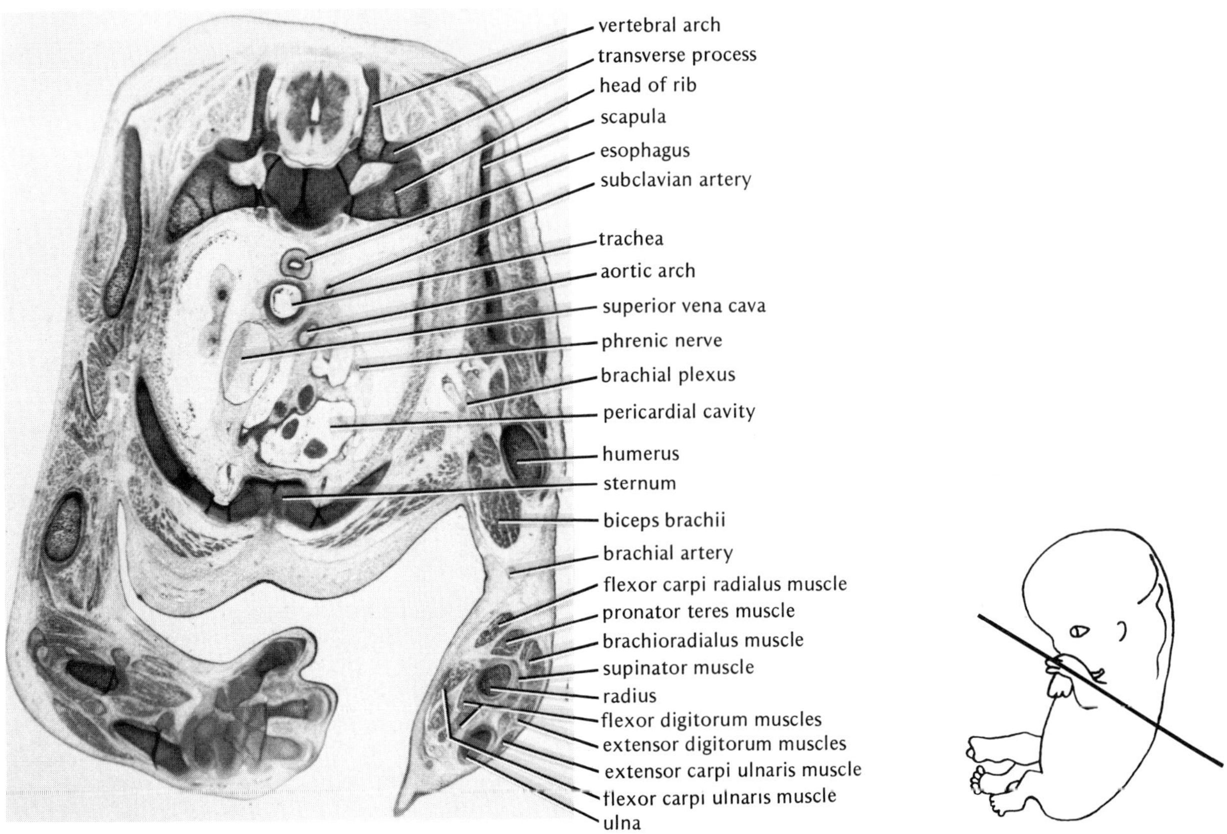

Figure 66 32-mm human embryo, transverse section at superior vena cava level

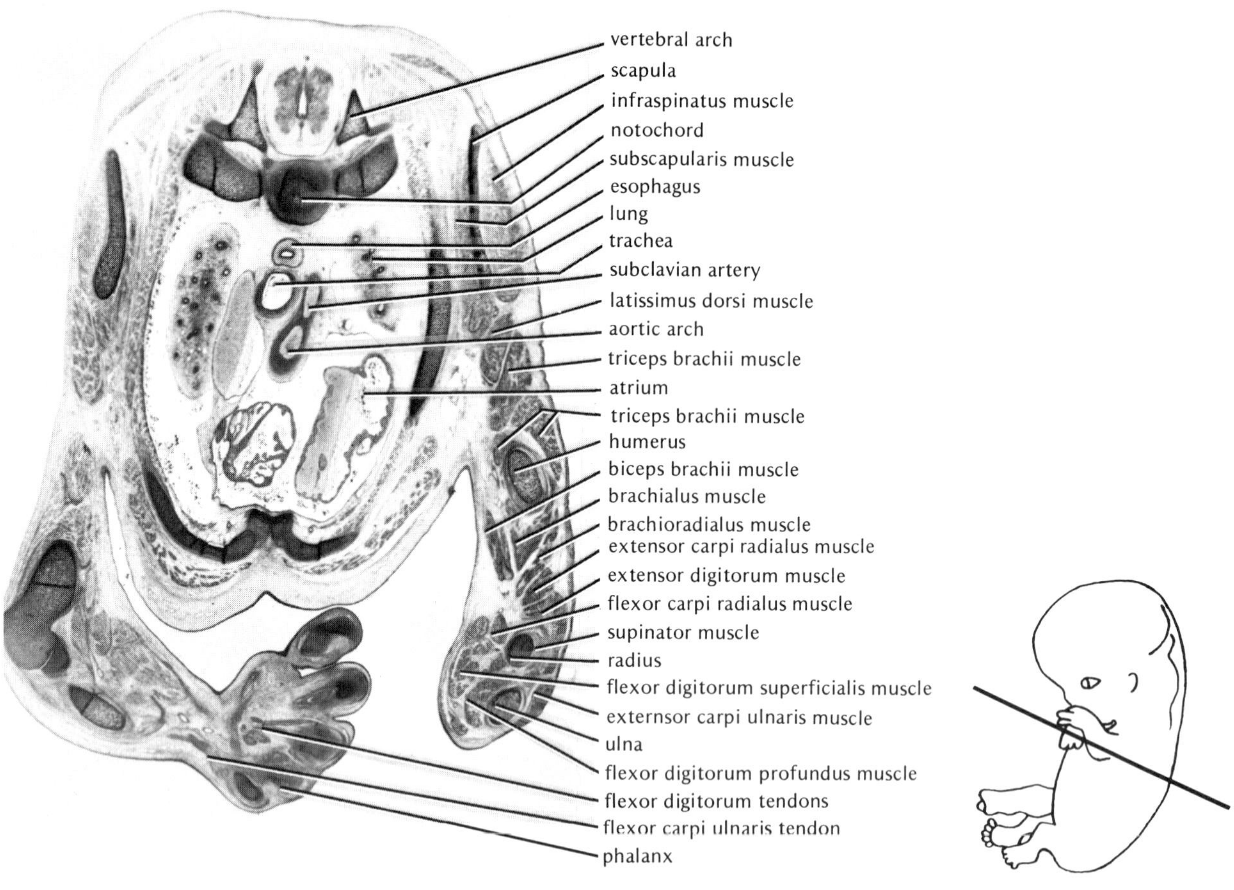

Figure 67 32-mm human embryo, transverse section at arm level

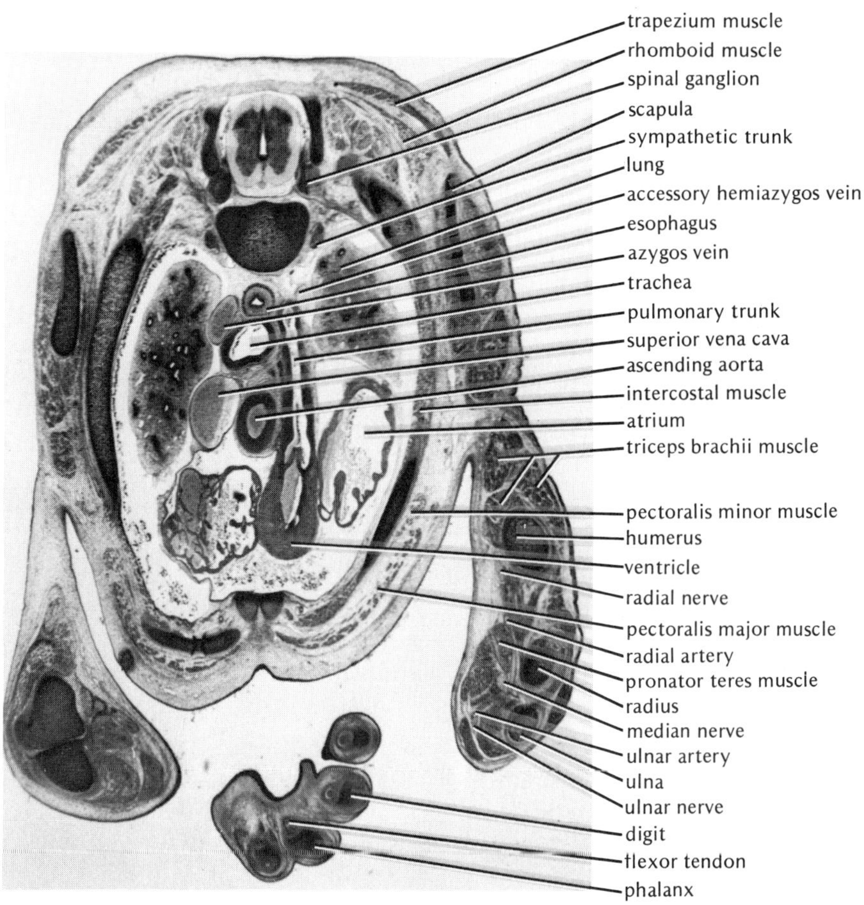

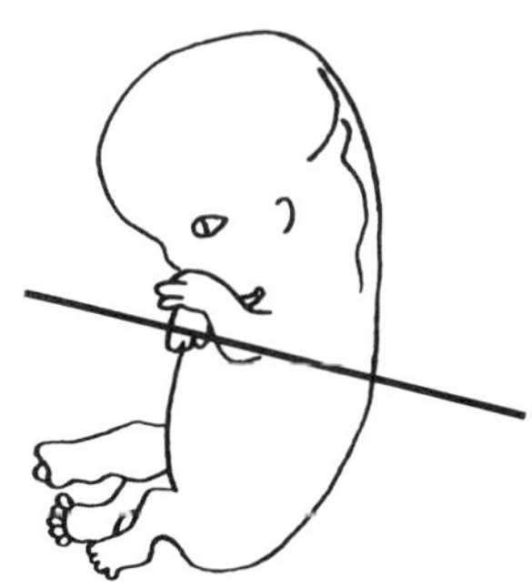

Figure 68 32-mm human embryo, transverse section at bulbus arteriosus level

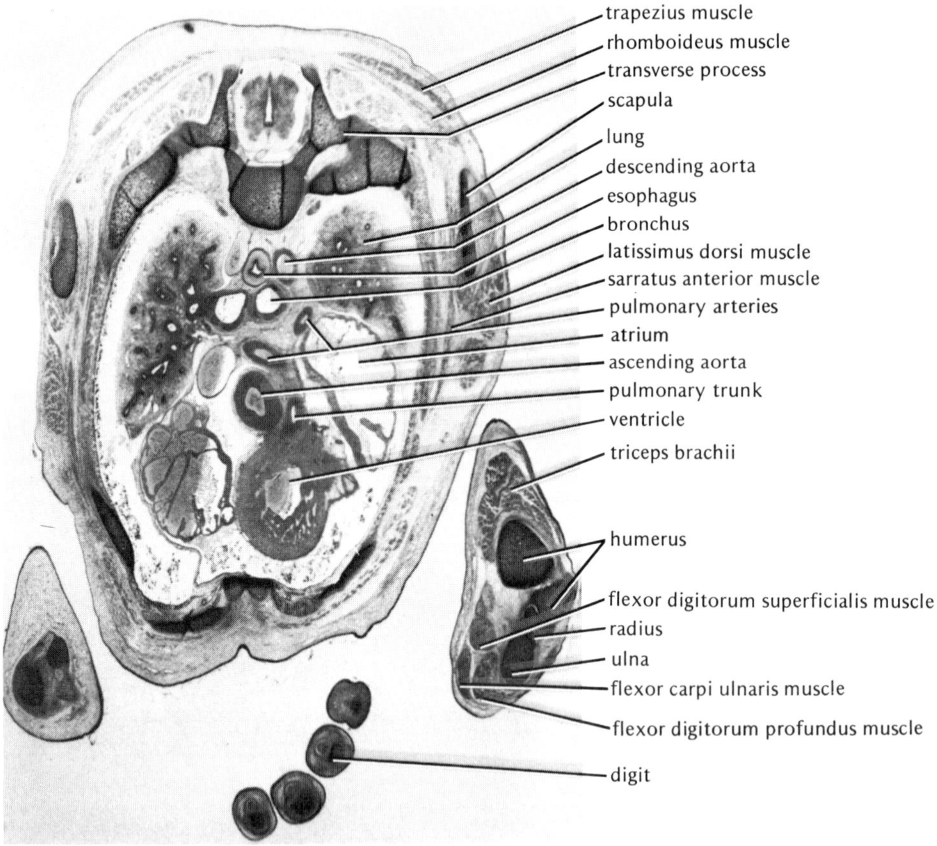

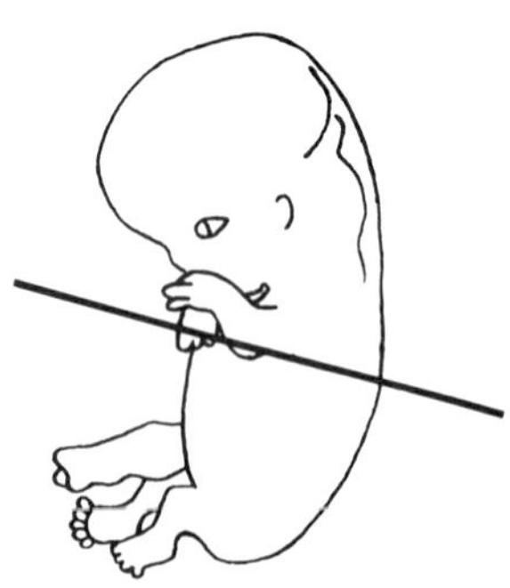

Figure 69 32-mm human embryo, transverse section at lung level

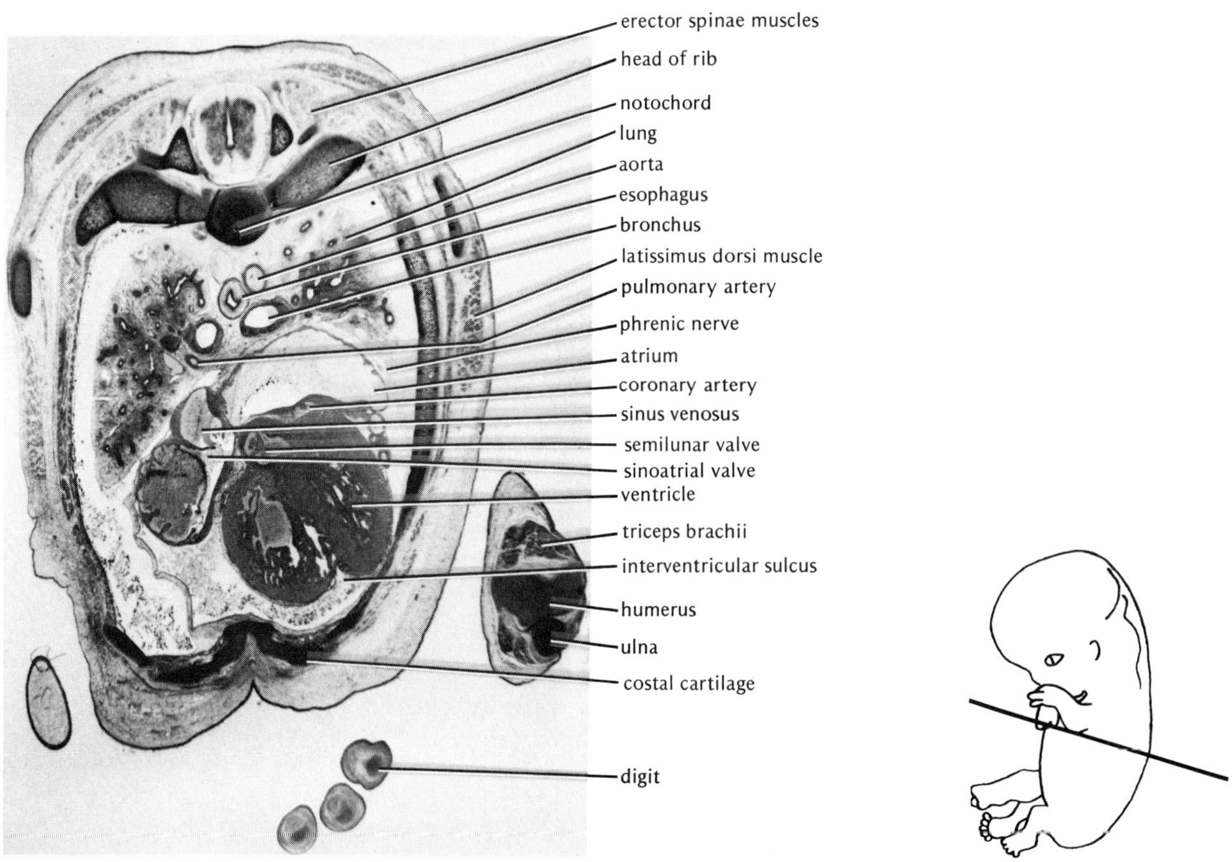

Figure 70 32-mm human embryo, transverse section at interatrial septa level

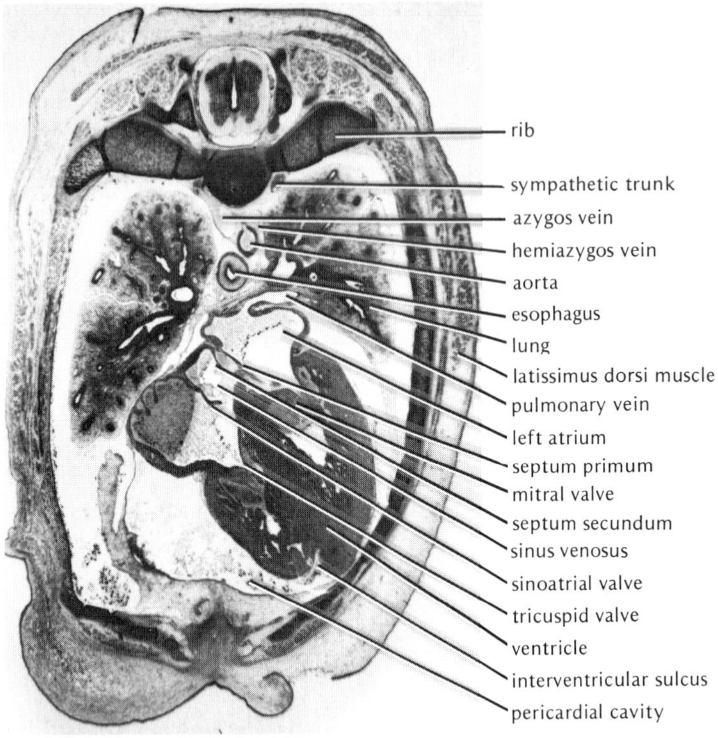

Figure 71 32-mm human embryo, transverse section at atrioventricular valve level

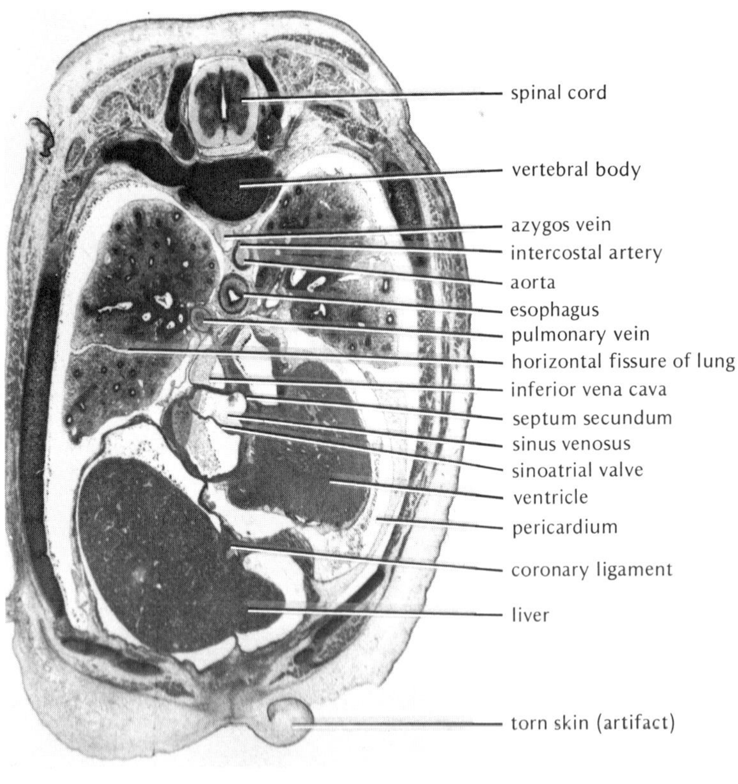

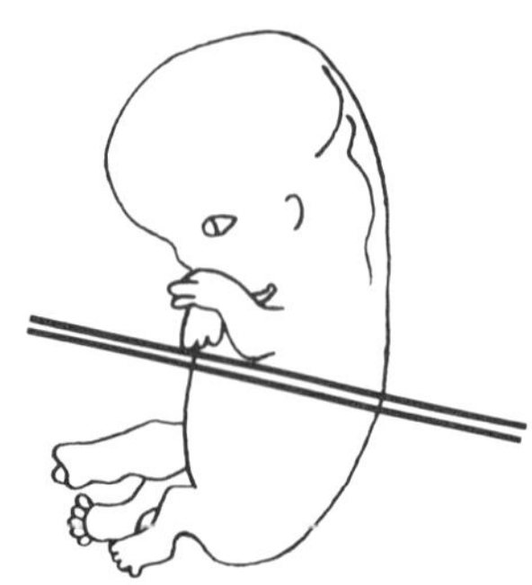

Figure 72 32-mm human embryo, transverse section at ventricle level

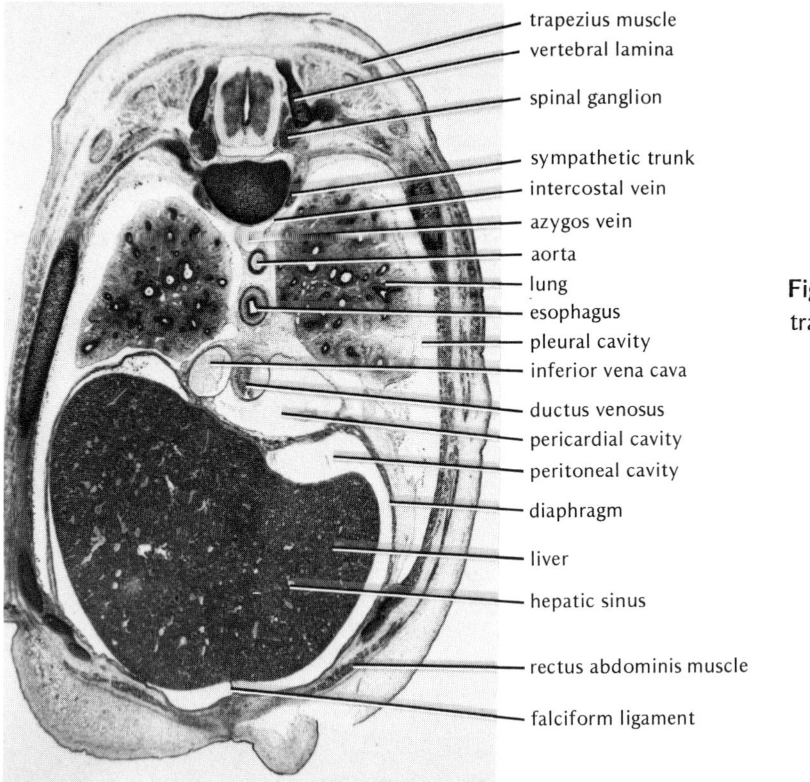

Figure 73 32-mm human embryo, transverse section at inferior vena cava level

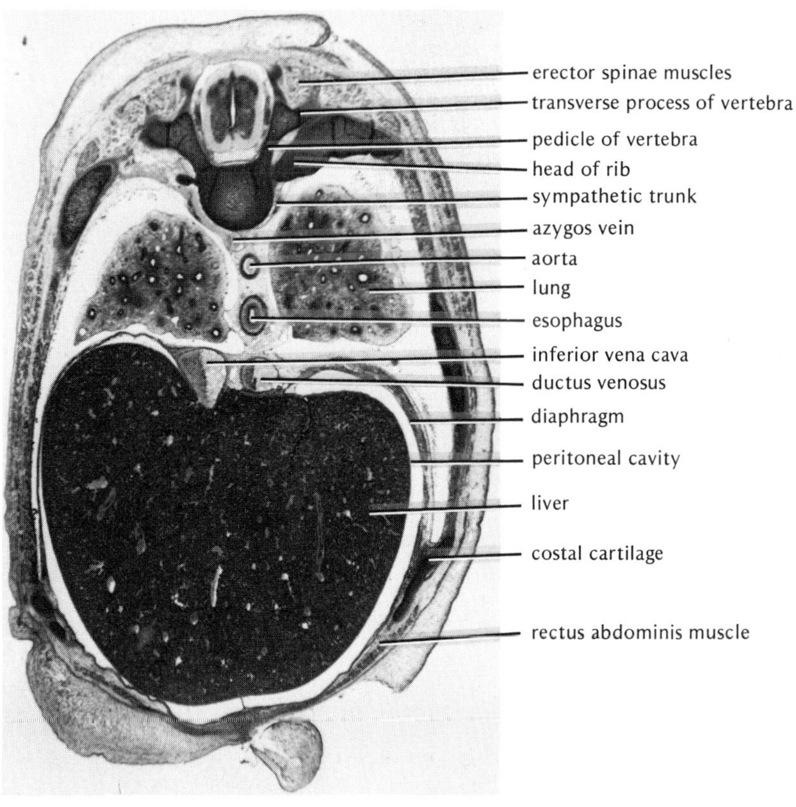

Figure 74 32-mm human embryo, transverse section at ductus venosus level

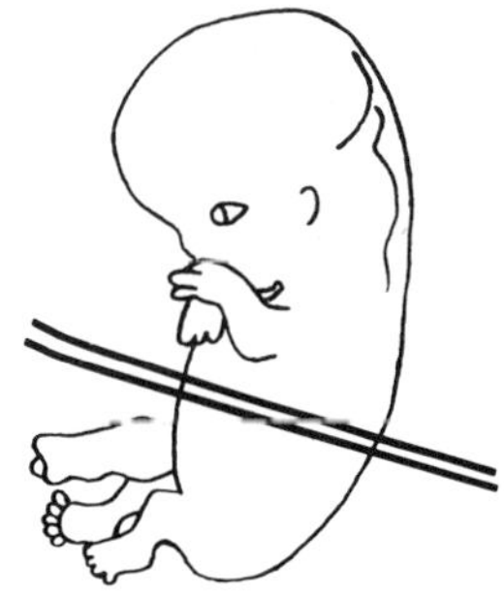

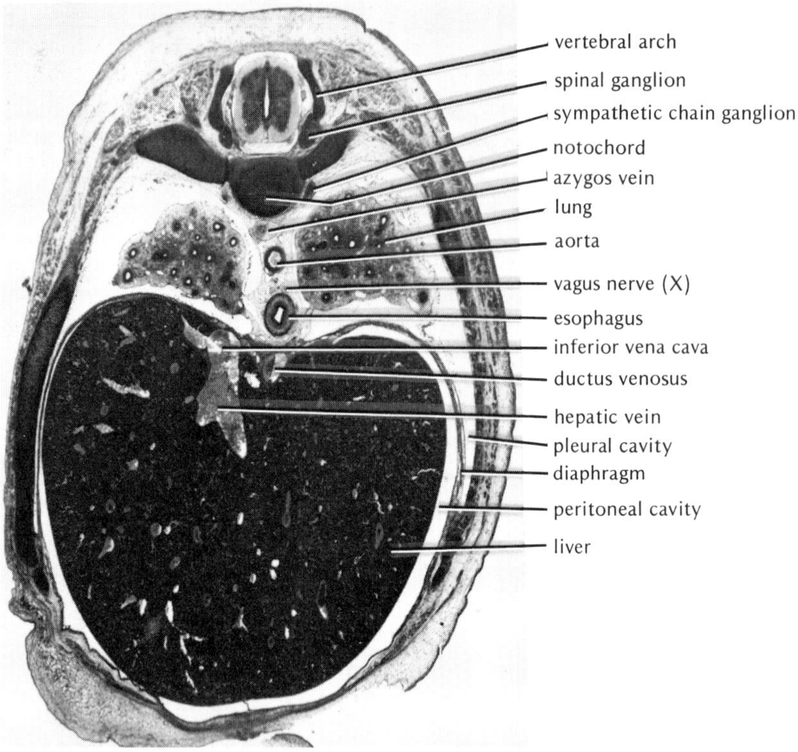

Figure 75 32-mm human embryo, transverse section at hepatic vein level

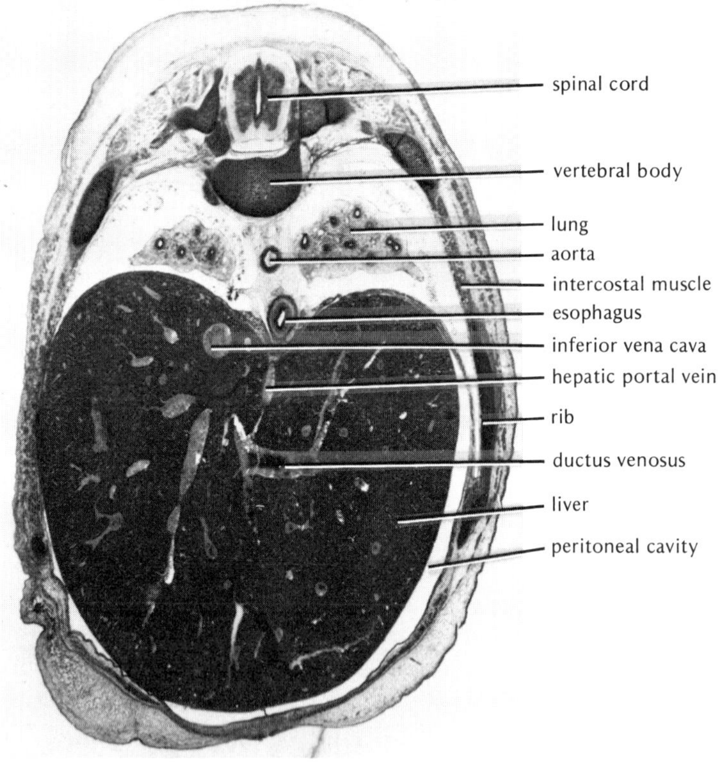

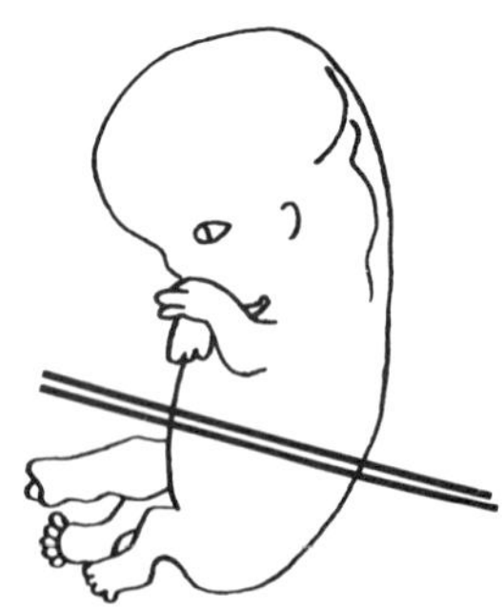

Figure 76 32-mm human embryo, transverse section at liver level

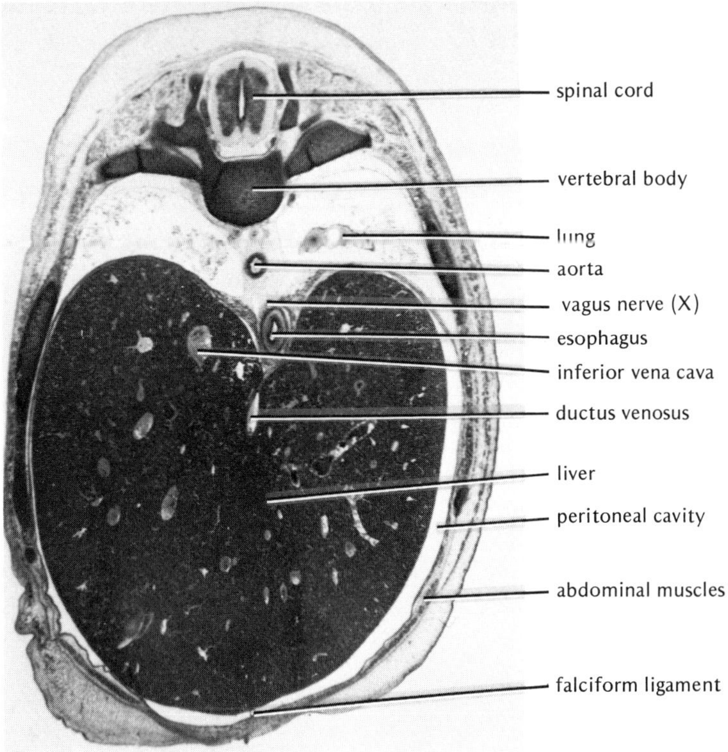

Figure 77 32-mm human embryo, transverse section at upper abdominal level

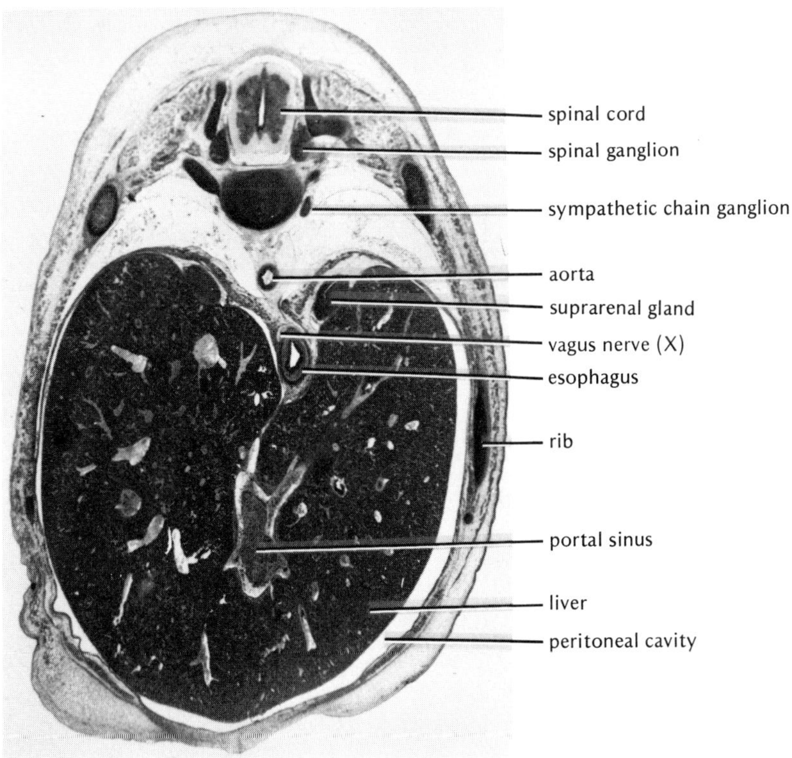

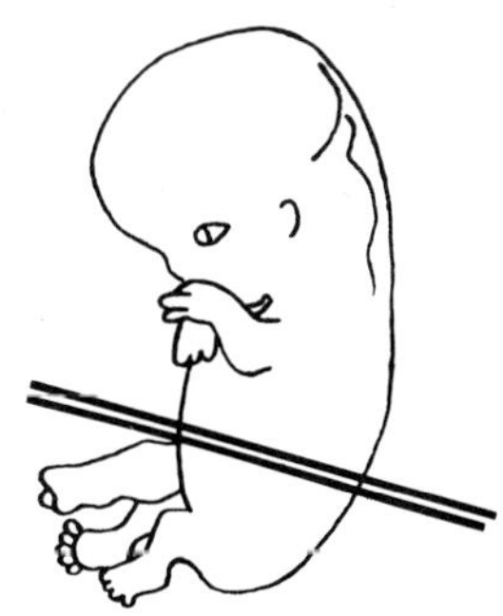

Figure 78 32-mm human embryo, transverse section at portal sinus level

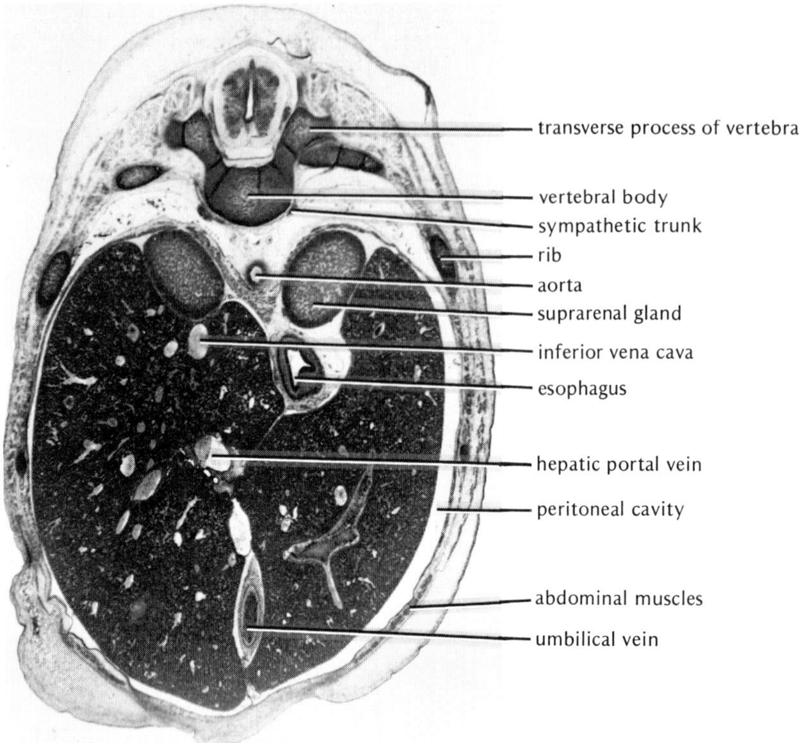

Figure 79 32-mm human embryo, transverse section at adrenal level

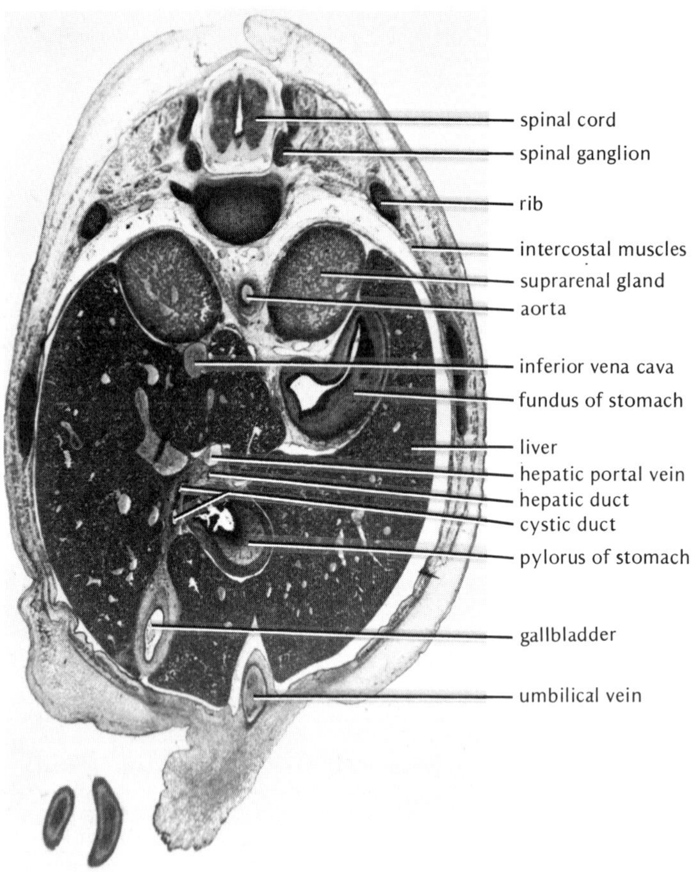

Figure 80 32-mm human embryo, transverse section at fundus level

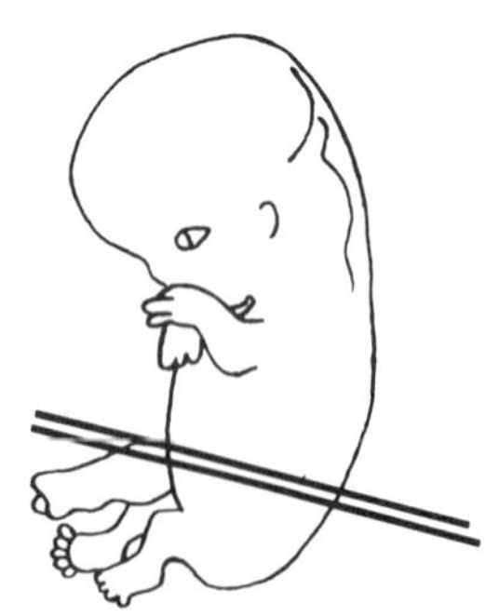

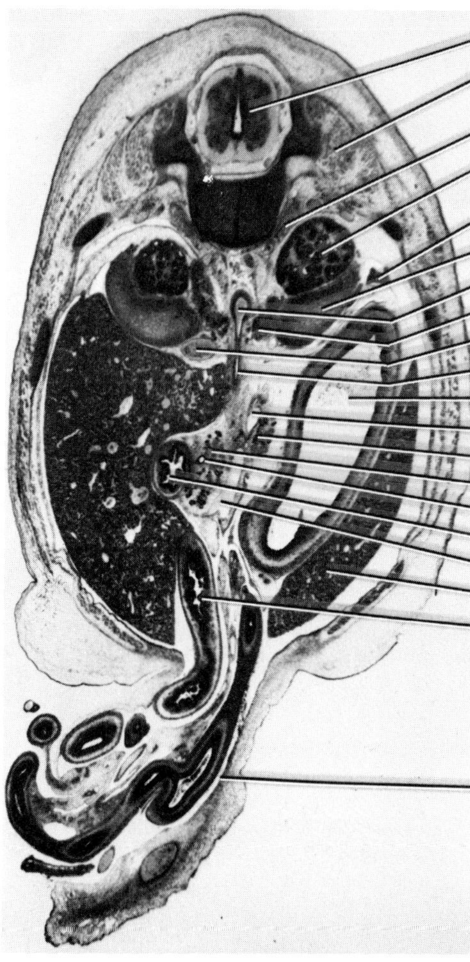

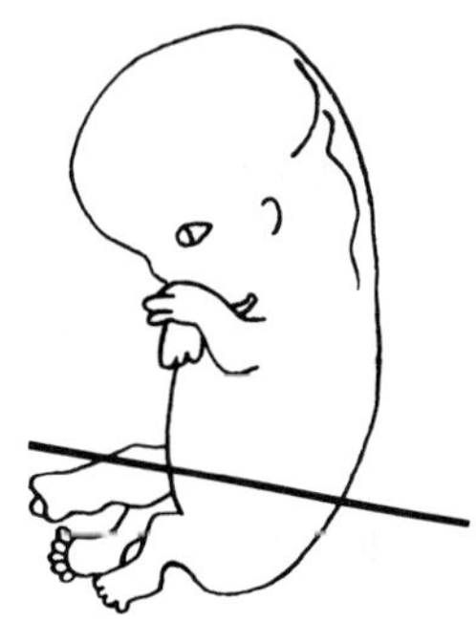

Figure 81 32-mm human embryo, transverse section at stomach level

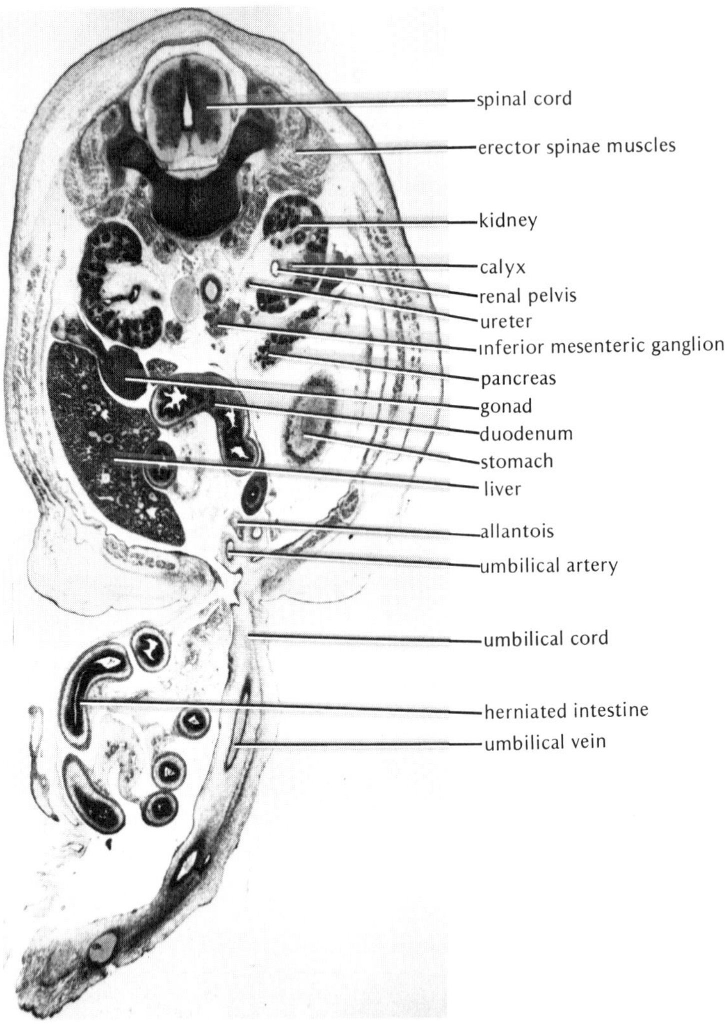

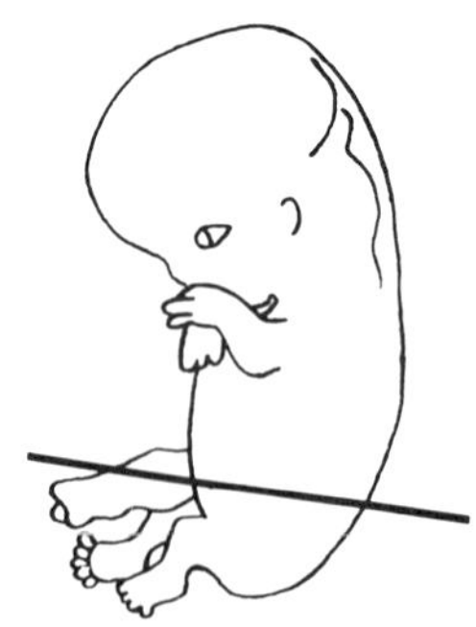

Figure 82 32-mm human embryo, transverse section at kidney level

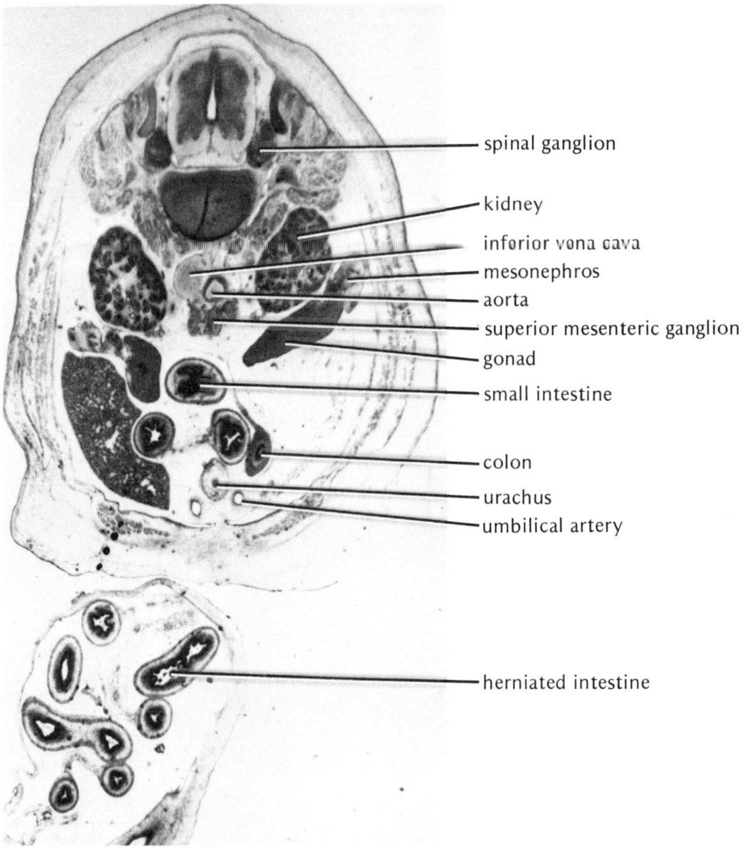

Figure 83 32-mm human embryo, transverse section at gonad level

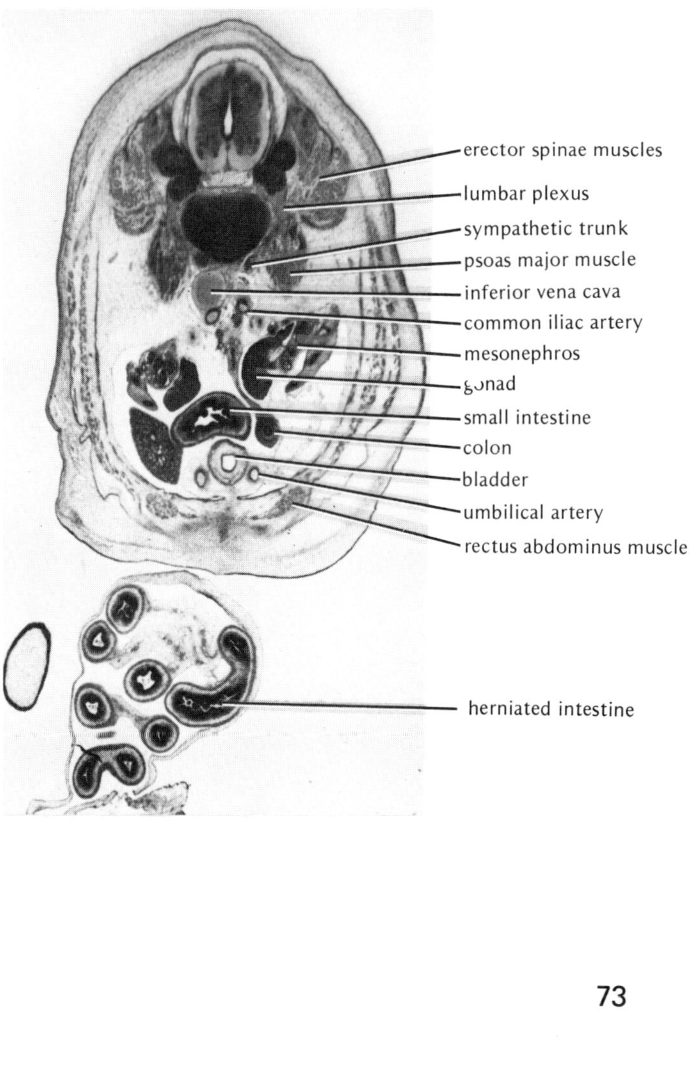

Figure 84 32-mm human embryo, transverse section at umbilical hernia level

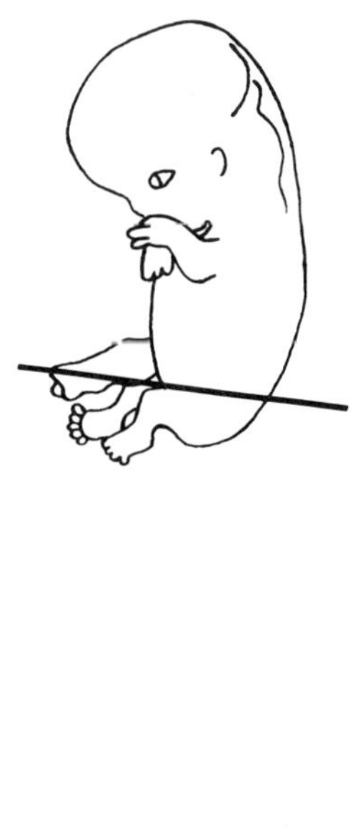

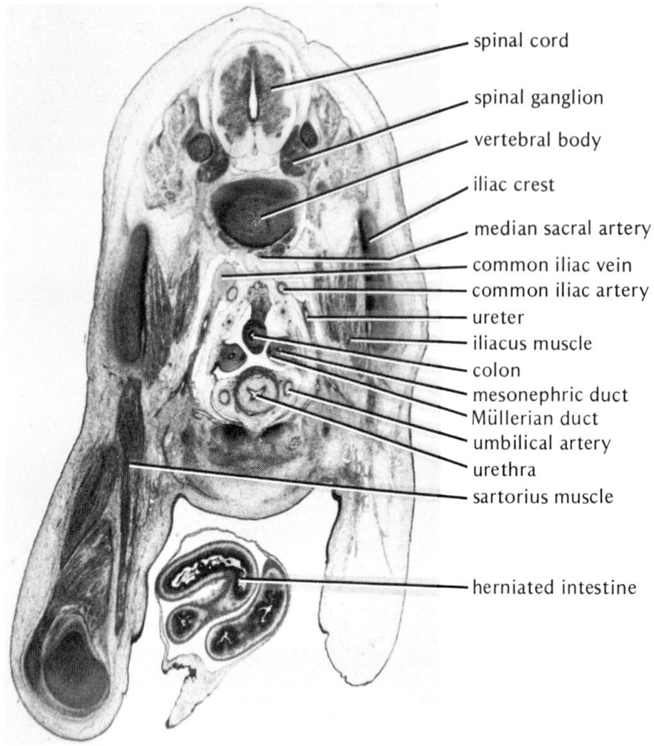

Figure 85 32-mm human embryo, transverse section at ilium level

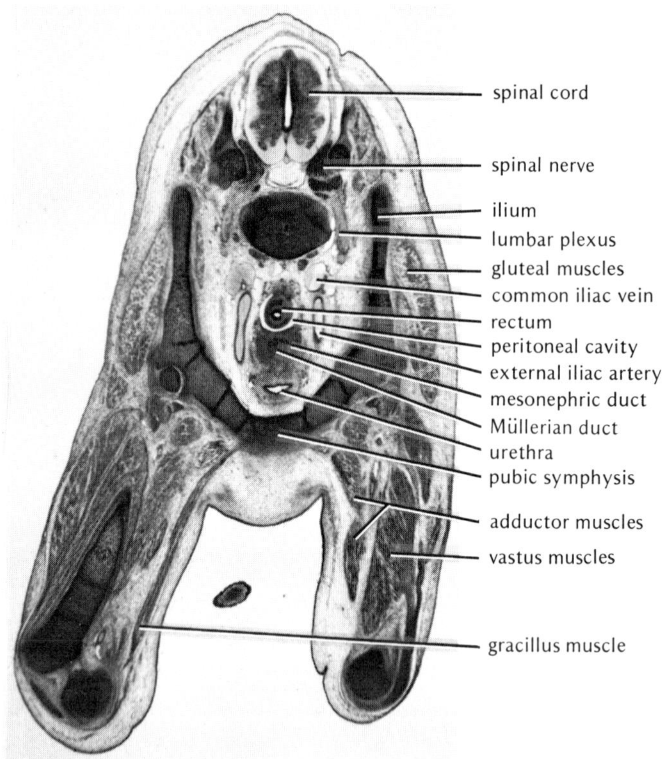

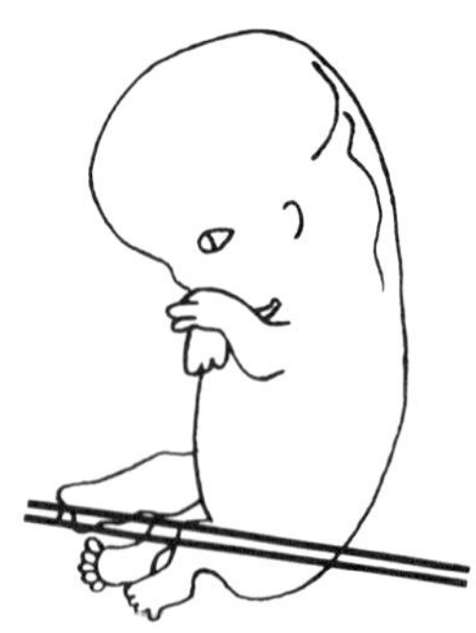

Figure 86 32-mm human embryo, transverse section at pubic symphysis level

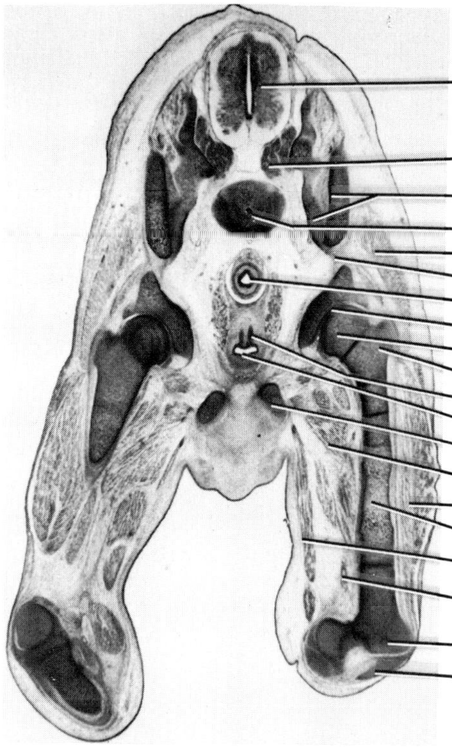

Figure 87 32-mm human embryo, transverse section at hip level

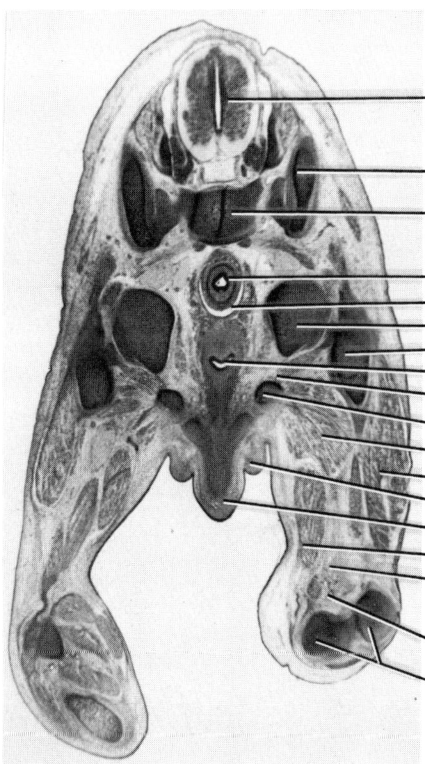

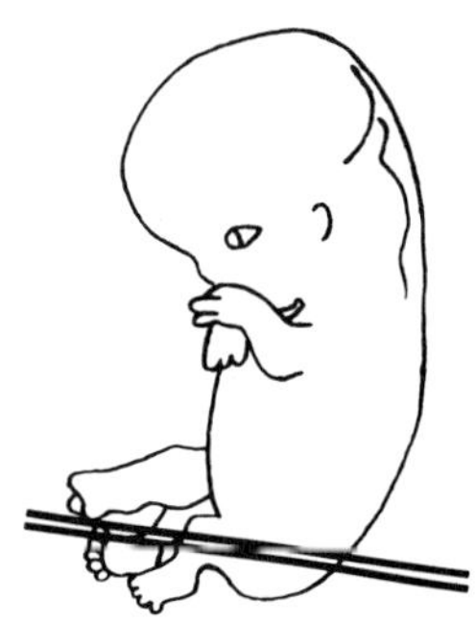

Figure 88 32-mm human embryo, transverse section at thigh level

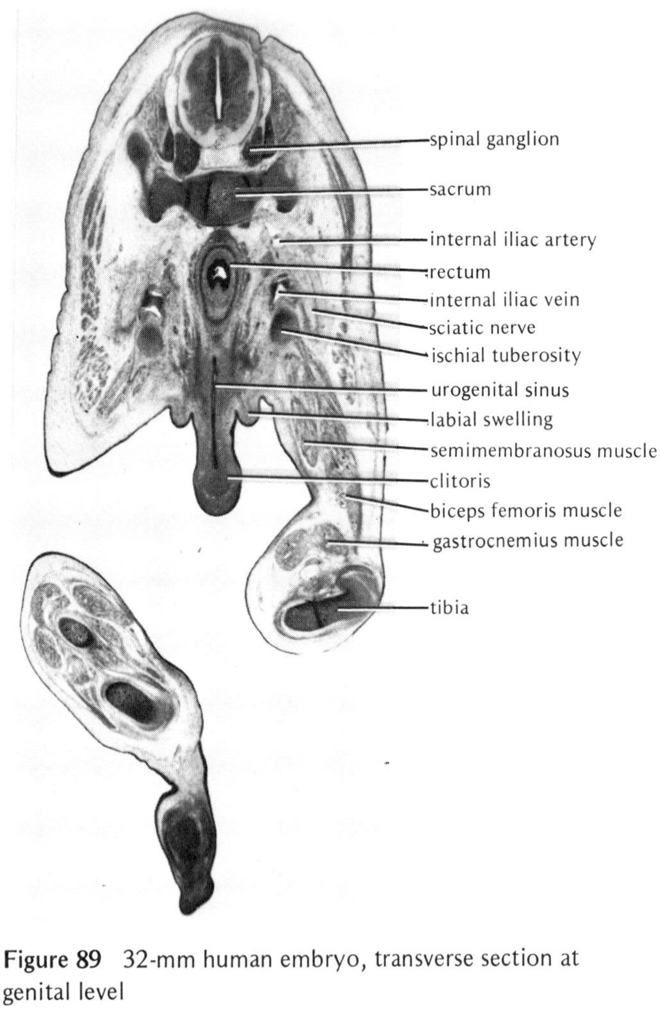

Figure 89 32-mm human embryo, transverse section at genital level

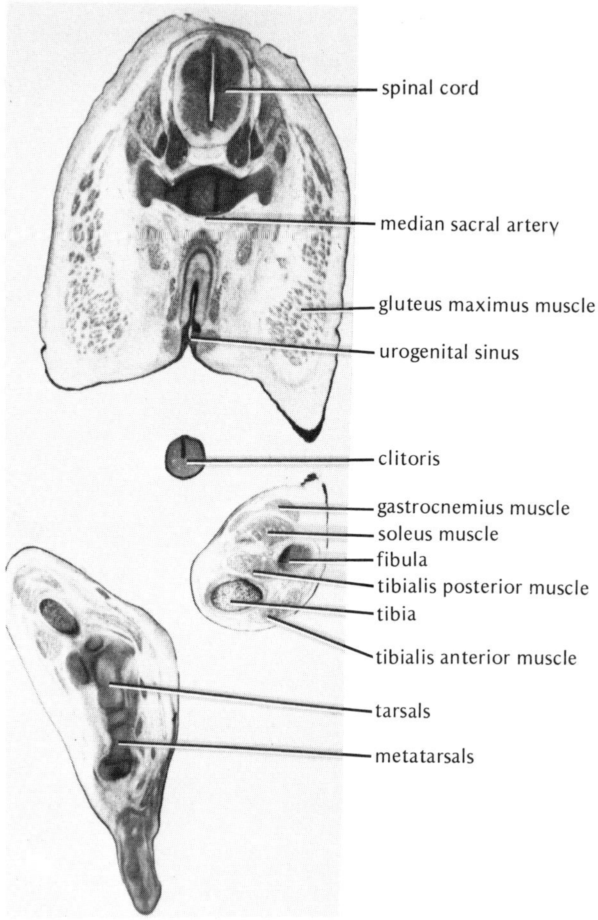

Figure 90 32-mm human embryo, transverse section at anal level

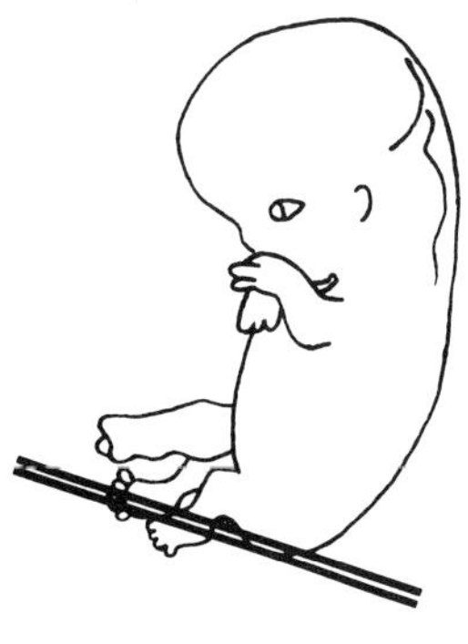

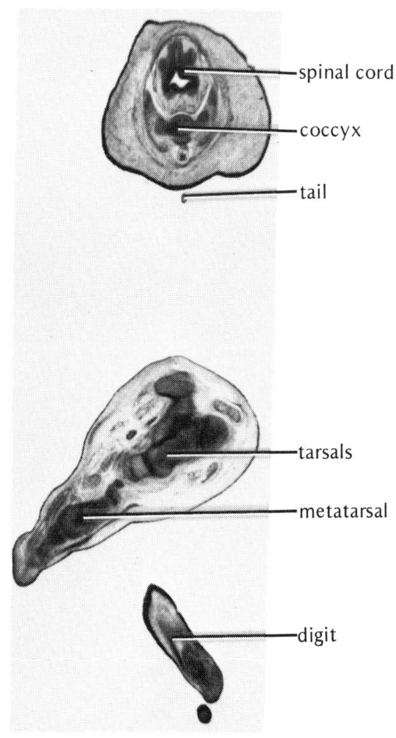

Figure 91 32-mm human embryo, transverse section at tail level

Recommended Reading

Arey, L.B. (1974): *Developmental Anatomy,* Saunders, Philadelphia, Pa.

Blechschmidt, E. (1961): *The Stages of Human Development Before Birth,* Saunders, Philadelphia, Pa.

Corliss, C.E. (1976): *Patten's Human Embryology,* McGraw-Hill, New York

Gasser, R. (1975): *Atlas of Human Embryos,* Harper & Row, Hagerstown, Md.

Hamilton, J.J., and Mossman, H.W. (1972): *Human Embryology,* Williams and Wilkins, Baltimore, Md.

Mathews, W.W. (1976): *Atlas of Descriptive Embryology,* Macmillan, New York

O'Rahilly, R. (1973): *Developmental Stages in Human Embryos. Part A: Embryos of the First Three Weeks, Stages One to Nine,* Academic Press, New York.

(1975): *A Color Atlas of Human Embryology,* Saunders, Philadelphia, Pa.

Streeter, G.L. (1951): *Developmental Horizons in Human Embryos. Age Groups 11 to 23,* Vol. 2, Carnegie Institute, Washington, D.C., Distributed by Academic Press, New York.

Index and Glossary

A

Abducens Nerve (VI): *Fig. 56;* (abduct: away); 6th cranial nerve to the lateral rectus muscle of the eye, therefore functions to abduct the eye; appears at about 30 days.

Allantois: *Figs. 31-34, 82*; (allanto-: sausage); membranous evagination of the hindgut region; appears at about 15 days and develops into the bladder and an extended tube, the urachus, which later becomes closed as a ligament of the adult bladder; if the lumen persists, it may result in a vesicoumbilical fistula.

Aorta: *Figs. 20, 42, 68-81, 83*; (aort-: lift up); the major vessel of the systemic arterial system; begins about 18 days as a ventral and dorsal aorta joined by six aortic arches ; the 4th left aortic arch persists between the ventral (ascending aorta) and dorsal (descending aorta).

Aortic Arch: *Figs. 18-20, 66, 67*; six pairs of arches which connect the dorsal and ventral aorta by 25 days; the 1st two pairs are lost, the 3rd pair becomes part of the internal carotid arteries, the 4th pair becomes the subclavian on the right and the arch of the aorta on the left, the 5th pair is lost, and the 6th pair gives rise to the pulmonary arteries.

Apical Ectodermal Ridge (AER): *Figs. 9, 23, 24*; a ridge of thickened, columnar epithelial cells at the apex of the developing limb bud, responsible for inducing elongation of the developing limb, first appears at about 25-30 days and disappears at about 40-42 days.

Arytenoid Swelling: *Figs. 7, 18, 61*; (arytenoid: ladle); a swelling whose cartilage will become a portion of the adult larynx, first appears at 42 days.

Atrium: *Figs. 8, 20, 21, 42, 43, 67-70*; (atrium: entrance hall); one of the 1st four, unpaired chambers of the embryonic heart appearing at about 22 days; divided into two chambers by formation of the septum primum (42 days) and septum secundum (55 days); as the atrium expands (40-50 days) the trunks of the inflow veins are absorbed and become the smooth wall of the adult atria, the embryonic atria are represented by the rough-walled portion of the adult atria and remain primarily as the auricles.

Auditory Nerve (VIII): *Figs. 14, 15, 39, 57;* a sensory cranial nerve carrying both hearing and balance sensations from the inner ear; first appears between 28 and 32 days.

Azygos Vein: *Figs. 68, 71-75*; (azygos: unpaired); vein in the upper abdomen and thorax; receives blood from the intercostal and hemiazygos veins and empties into the superior vena cava, develops as the right supracardinal at about 45-50 days.

B

Basilar Artery: *Figs. 15, 38, 39, 53-57*; artery that runs along the base of the brain from the vertebral arteries to the circle of Willis; develops by about 32 days.

Bowman's Capsule: *Figs. 30, 31*; (See Glomerulus.)

Brachial Artery: *Fig. 66*; major artery supplying the arm, develops from the central arm bud artery by about 45 days.

Brachial Plexus: *Figs. 23, 42, 66*; (plexus: braid); the network of nerves supplying the arm; present in the embryo by about 38 days.

Brachiocephalic Artery: *Fig. 65*; an artery in the right upper thorax that carries blood from the arch of the aorta and divides into subclavian and common carotid arteries; develops from the 4th aortic arch by about 32 days.

Brachiocephalic Vein: *Fig. 41*; paired veins arising from the jugular and subclavian veins on each side of the upper thorax and joining the opposite member of the pair to form the superior vena cava; the left develops from a vein connecting the left and right anterior cardinal veins and the right from the right anterior cardinal vein; both develop by about 50 days.

Bronchus: *Figs. 22, 23, 43, 69, 70*; the two major tubes leading from the trachea, one leading to each lung; they first appear as bronchial buds from the trachea at about 28 days.

Bulbus Cordis: *Fig. 20*; the most anterior of the four primitive heart chambers appearing at about 22 days and connecting the ventricle with the ventral aorta (truncus arteriosus); a large portion of the bulbus is absorbed into the right atrium as the conus arteriosus; the remainder together with the truncus arteriosus is divided by a spiral septum into ascending aorta and pulmonary trunk, at about 35 days.

C

Cardinal Veins: Common: *Figs. 21, 22, 42, 43*; **Anterior:** *Figs. 6, 15, 16, 18, 20*; **Posterior:** *Figs. 9, 23-33*; (cardinal: principal); the 1st major veins of the embryo bringing all the blood back to the heart except that from the placenta; the anterior and posterior cardinals on each side join to form short common cardinals that drain into either side of the sinus venosus; they develop by 28 days; the posterior cardinals almost completely disappear by 50 days; the anterior cardinals persist as the internal jugular veins throughout life.

Carotid Arteries: Common: *Figs. 63-65*; **Internal:** *Figs. 6, 17, 39, 40, 57-62*; (carotid: put to sleep); the common carotid arteries carry blood from the brachiocephalic artery (right) and the arch of the aorta (left); both arteries pass up the side of the neck and divide into the internal and external carotid; the internal carotids pass into the interior of the skull to supply the brain, while the external carotids serve the face and neck; the internal carotids are extensions of the dorsal aorta appearing by about 22 days; the external carotids develop as extensions of the ventral aorta by 30 days, and the common carotids develop from the ventral aorta between the 3rd and 4th aortic arch by about 30 days.

Caudate Nucleus: *Figs. 54-58*; (caudate: tail); one of the major nuclei in each hemisphere of the brain; they form the interior-lateral wall of the lateral ventricles; together with the lentiform nuclei they make up the corpora striata; appear by about 35 days.

Cerebellum: *Figs. 51-54*; the smaller and more posterior of the two major brain masses; first develops as cerebellar plates on the lateral walls of the metencephalon at about 40 days.

Choroid Plexus: *Figs. 53-57*; (choroid: membrane); a lacy membrane network filled with blood vessels and located in the roof of each cerebral ventricle, produces the cerebral spinal fluid as a blood extract; they appear some time around 40-45 days.

Clitoris: *Figs. 88-90*; the erectile portion of the female external genitalia, the clitoris in the female or the comparable penis in the male appears as a genital tubercle by about 26 days; by 70 days the sex can be determined by the external genitalia; the relative changes progress until about 90–100 days, when the definitive adult characteristics can be recognized.

Cloaca: *Fig. 34*; (cloaca: sewer); the primitive common ending of the allantois, urogenital, and digestive tubes; the cloaca develops by about 25 days and is separated into urogenital sinus and rectum by 34 days.

Colon: *Figs. 30, 83–85*; large intestine; the colon can first be recognized as a distinct region of the hind gut by the formation of the cecum and separation of the cloaca into urogenital sinus and rectum at about 35 days, the cecal end remains outside the body proper as a part of the umbilical hernia until about 60 days, when it is drawn into the abdomen and takes up its definitive adult position.

Common Bile Duct: *Fig. 81*; a duct formed by the union of the hepatic and cystic ducts, which joins the pancreatic duct to empty into the duodenum at the ampulla of Vater; develops at about 30 days.

Cornea: *Figs. 40, 60*; (cornea: horny); the rather thick transparent outer layer of the anterior portion of the eye; first develops about 40 days.

Coronary Artery: *Fig. 70*; (coronary: crown); arteries encircling and feeding the tissue of the heart; they begin to appear by 50 days.

Coronary Ligament: *Fig. 72*; (coronary: crown); a circular ligament between the liver and diaphragm; it begins to develop at about 35 days as the liver moves out of the transverse septum while the latter is developing into the diaphragm; the coronary ligament surrounds the bare area of the liver.

Corpus Striatum: *Fig. 56*; (striated body). (See caudate and lentiform nuclei.)

Cystic Duct: *Fig. 80*; the duct leading from the gall bladder to the common bile duct; develops at about 30 days.

D

Diaphragm: *Figs. 73–75*; the muscular, transverse separation between the abdominal and thoracic cavities, composed of several interconnecting membranes; the midventral portion develops first as the transverse septum by about 25 days; this septum separates the pericardial from the peritoneal cavities; the pleural cavities are later separated from the peritoneal cavity by the development of the pleuroperitoneal membrane at about 40 days.

Diencephalon: *Figs. 6, 16–19, 38, 50*; (dien: through; cephalon: brain); the more posterior of the two portions of the forebrain consisting of the 2nd and 3rd neuromeres; the diencephalon can first be identified at 21 days; the optic cups develop from the diencephalon at about 25 days; the diencephalon becomes the thalamus, subthalamus, the hypothalamus by about 40 days.

Dorsal Aorta: *Figs. 6–10, 18–35, 42–46*; begins as paired dorsal arteries which develop by 20 days; receives blood from the ventral aorta through the aortic arches; becomes well defined as a single descending aorta by 35–40 days.

Dorsal Mesentery: *Figs. 9, 10, 31–33*; (mesentery: middle intestine); a sheet of serous tissue connecting much of the gastrointestinal tract to the dorsal surface of the abdominal cavity; it is a double membrane created by a continuation of the visceral and parietal peritoneum.

Dorsal Mesogaster: *Figs. 25-27*; (greater omentum); the dorsal mesentery of the stomach; develops into the greater omentum with the rotation of the stomach at about 45–50 days.

Dorsal Root: *Fig. 22*; division of the spinal nerve which carries afferent sensory impulses to the spinal cord; contains the dorsal root or spinal ganglion; develops from neural crest cells by 35 days.

Ductus Venosus: *Figs. 24, 25, 26, 73-77*; a temporary vein passing through the liver to shunt most of blood from the umbilical vein directly to the heart without filtering through the sinusoids; develops by 35 days and remains until birth.

Duodenum: *Figs. 27, 81, 82*; (duodenum: twelve); first 12 inches of the adult GI tract after the stomach; receives the common bile and pancreatic ducts; can be identified as soon as the stomach swelling first appears at 30 days.

Dura Mater: *Figs. 51, 58*; (dura mater: tough mother); the outer covering of the central nervous system; contains the blood sinuses.

E

Embryonic Disc: *Fig. 3*; a double layer of cells which develops between the amniotic and yolk sac cavities at about 7 days from which the embryo proper develops; the designation is discontinued after neurulation begins at about 18 days.

Enamel Organ: *Figs. 61, 62*; early organizer to the tooth enamel, responsible for tooth shape; they first appear on the labial side of the dental lamina at about 60 days and then degenerate when the enamel is complete, just before birth.

Endocardial Cushion: *Fig. 22*; enlarged portions of the primitive heart wall between the atrium and ventricle forming a temporary AV "valve" until the true valves develop; it first appears at about 22 days and becomes the focal point for the development of the heart septa and AV valves.

Endolymphatic Duct: *Figs. 13, 14, 56, 57*; a duct leading from the vestibule of the inner ear superiorly to a blind endolymphatic sac; begins to develop as an evagination from the vestibule at about 30 days.

Ependymal Layer: *Figs. 12, 16, 49, 50*; (ependyma: clothing); the innermost of the three layers of nervous tissue of the central nervous system; first appears at about 25–30 days during neural tube closure; it is the layer of active cell proliferation and is the thickest layer during early development (at 50 days accounts for about 60 percent of the total thickness); by 70 days it accounts for only 25 percent of the total thickness; it continues to decrease in relative thickness until it becomes a single-celled membrane surrounding the internal ventricles and canals of the CNS and remains as such through life.

Epiphysis: *Figs. 19, 50*; (pineal body); (epiphysis: grow upon); (pineal: pine cone); a dorsal projection from the diencephalon; its function is unclear, but in lower animals it appears to function in photoperiodicity and may also respond to light-dark cycles in humans as well; it appears as an evagination at about 45–50 days, begins to degenerate at about age 7 and in the adult remains largely as connective tissue.

Esophagus: *Figs. 8, 20-24, 41-45, 63-79*;

that portion of the GI tract between the pharynx and stomach; it can be identified by 30 days when the stomach swelling first appears.

External Ear: *Figs. 40, 58, 59*; (auricle); begins to develop from portions of the 1st and 2nd branchial arches at about 35 days, and by 40-45 days exhibits distinct adult characteristics.

Eye: *Fig. 61*; the eye can first be seen as an optic field on the lateral wall of the developing diencephalon by 22 days; by 25 days optic vesicles have evaginated remaining attached to the diencephalon by an optic stalk (future optic nerve); an invagination of the lateral wall of the optic vesicle at about 30 days results in the formation of the double-walled optic cup (future retina); at this same time the optic cup reaches the lateral head ectoderm and induces it to invaginate as a lens vesicle; this vesicle pinches off at about 35 days and becomes the solid lens by 45 days; the ectoderm remaining behind as the lens pinches off becomes the cornea of the eye.

Eyelid: *Figs. 60, 61*; begin to appear as folds adjacent to the eye at 55 days, then grow together and fuse by 70 days and reopen at about seven or eight months.

F

Facial Nerve (VII): *Figs. 16, 17, 39, 56, 57*; motor nerve of the face; develops at about 30 days.

Falciform Ligament: *Figs. 25, 44, 73, 77*; (falciform: sickle-shaped); mesentery connecting the diaphragm and anterior abdominal wall to the liver; the round ligament is in the free edge of this ligament (see umbilical vein); it begins to develop at about 30-35 days.

Falx Cerebri: *Figs. 40, 52, 53, 55-58*; (falx: sickle); dural reflection in the longitudinal fissure between the lateral hemispheres.

Foramen Primum: *Fig. 22*; the 1st ostium connecting the two atria; the atria are continuous from their beginning, but the foramen primum is first defined when the septum primum begins to develop at about 30 days; it disappears as the septum primum is completed, at about 40 days.

Fourth Ventricle: *Figs. 12, 14, 38, 51-55*; the central cavity of the metencephalon in the region of the cerebellum; it appears with closure of the neural folds at 22 days.

G

Gallbladder: *Figs. 27, 80*; an endodermal sac for the storage of bile; it develops as an evagination of the liver diverticulum at about 30 days, and elongates, producing a cystic duct at about 38 days which, together with the hepatic duct, forms the common bile duct.

Geniculate Ganglion (VII): *Fig. 15*; ganglion of the facial nerve.

Genital Ridge: *Figs. 29-33*; a ridge of tissue which develops along the ventral edge of the mesonephros appearing at about 35 days; they are seeded with germ cells which migrate into the region from the yolk sac; this tissue develops into the gonad.

Genital Tubercle: *Figs. 32, 46*; the primordia of the female clitoris or male penis; first appears at about 26 days as two bilateral swellings, which later fuse; see clitoris.

Glomerulus: *Figs. 26, 29*; a capillary plexus inside a Bowman's capsule; site of urine filtration; a mesonephric group first develops at about 30–35 days and degenerates at about 60–70 days; the metanephric group develops at about 50-60 days.

Glossopharyngeal Nerve (IX): *Figs. 14, 15, 17, 59*; nerve supplying motor and sensory fibers to the tongue and pharynx; appears at about 30 days.

Gonad: *Figs. 82–84*; ovary in the female, testis in the male; develops from the genital ridge.

H

Hemiazygos Vein: *Figs. 44, 71*; vein which develops from the left supracardinal vein, which originates at about 45–50 days and empties into the azygos vein.

Hepatic Duct: *Fig. 80*; duct connecting the liver diverticulum to the GI tract from which it evaginated at about 30 days; joins the cystic duct to form the common bile duct.

Hepatic Portal Vein: *Figs. 27, 76, 79-81;* vein carrying blood from most of the abdominal viscera to the liver; develops from the left omphalomesenteric vein.

Hepatic Vein: *Fig. 75*; veins carrying blood from the liver to the inferior vena cava; develop from the omphalomesenteric veins.

Hippocampus: *Figs. 53, 54*; (hippocampus: sea horse); brain structure in the medial wall of the lateral ventricles.

Hyoid Arch: *Figs. 7, 18*; (Hyoid: V-shaped); primordia of the hyoid bone.

Hypoglossal Nerve (XII): *Figs. 14, 16-19, 39, 40, 58*; nerve to the intrinsic tongue muscles; develops at about 30 days.

Hypophysis: *Figs. 16, 39, 58*; (hypophysis: undergrowth); pituitary; compound gland developed as evaginations of the mouth (Rathke's pocket) and brain (infundibulum) at about 30 days.

Hypothalamus: *Figs. 18, 38, 39, 52-55, 57, 58*; an inferior projection of the diencephalon attaching the hypophysis to the brain.

I

Iliac Arteries and Veins: Common: *Figs. 85-86*; **Internal:** *Fig. 89*; **External:** *Fig. 86*; vessels between the abdominal aorta or inferior vena cava and internal pelvis (internal) or femorals of the leg (external); they develop at about 38 days.

Inferior Vena Cava: *Figs. 23, 27*; major vein returning blood from the lower portions of the body; develops from anastomosis between the subcardinal veins at about 40 days.

Infundibulum: *Figs. 6, 17*; (infundibulum: funnel); inferior projection of the diencephalon, connecting the hypothalamus and the hypophysis.

Inner Cell Mass: *Fig. 2*; a thickened portion of the blastocyst that will give rise to the embryo proper; appears at 4–5 days and is replaced by the embryonic disc at about 7 days.

Internal Jugular Vein: *Figs. 39, 40, 58, 60, 62-64*; (jugular: throat); the vein draining blood from the cranial cavity; develops as the primary head vein by 25 days.

Intersegmental Artery: *Figs. 24-27, 33-35, 72, 73*; vessels carrying blood from the dorsal aorta dorsally between the somites to the sides of the neural tube; they arise at about 20 days; the anterior segments combine to form the vertebral arteries by 27 days; the posterior segments remain as the intercostal and intervertebrals.

Interventricular Foramen: (heart); *Figs. 8, 22*; connecting hole between left and right ventricles; the ventricles are first one common cavity and the interventricular foramen is first defined as the interventricular septum develops by 35 days; the foramen is closed by the membranous portion of the interventricular septum by 40 days.

Interventricular Foramen: (of Monro, brain); *Figs. 20, 39, 56, 57*; connection between the lateral ventricles and the 3rd ventricle; appears as the lateral ventricles expand at about 35 days.

Interventricular Septum: *Fig. 22*; wall between the heart ventricles, most of which is muscular; begins to develop by 30 days and is completed by closure of a small superior membranous portion by 40 days.

Intestine: *Figs. 9, 10, 28-33, 46, 81-85;* inferior portion of the GI tract developing by closure of the yolk sac by 30 days, becomes coiled and extends into the umbilical cord by 35 days, and is retracted into the abdomen by 65 days.

J

Jugular Ganglion (X): *Fig. 14*; superior ganglion of the vagus nerve.

Jugular Lymph Sac: *Fig. 60-62*; cavernous opening in the mesenchyme of the neck; develops at 35 days from the local vascular plexus; the sac expands into a major neck structure by 50 days; by 60 days the sacs have developed into the definitive adult lymphatic pattern.

K

Kidney: *Figs. 81-83*; (metanephros); first appears as an evagination from the mesonephric duct by 30 days (pelvis), which extends and induces local mesoderm to develop into the cortex; takes over as the functional kidney by 50 days.

L

Lamina Terminalis: *Figs. 39, 52, 53, 55*; most anterior portion of the medial telencephalon; remains relatively stationary as the lateral ventricles expand anteriorly past it.

Larynx: *Figs. 18, 61-63*; voice box; most superior portion of the respiratory tract; begins to develop as the trachea evaginates from the GI tract at 28 days.

Lateral Ventricle: *Figs. 20, 22, 39, 40, 51-56, 58, 59*; lateral expansions of the telencephalon developing from the telencephalic vesicle, which appear by 30 days.

Lens: *Figs. 6, 18, 40, 60*; the refractive portion of the eye responsible for focusing light; develops from head ectoderm as a placode at 28 days; forms into a vesicle by 30 days; pinches off at 32 days; fills in to form a solid sphere by 40 days.

Lentiform Nucleus: *Figs. 56, 57*; (lentiform: lens shaped); nucleus of the central core of the cerebrum; with the caudate nucleus comprises the corpus striatum; develops at about 35 days.

Limb Bud: *Figs. 9, 10, 24, 26, 27, 34, 35, 41, 42, 46, 47*; future arm or leg; the arm buds appear at about 26 days, the leg buds at about 29 days; the buds elongate and the distal ends expand into hand plates (35 days); digits develop at about 40–45 days.

Limiting Sulcus: (sulcus: ditch); *Figs. 13, 14, 23, 38, 52*; longitudinal groove on the inner surface of the lateral walls of the neural tube; separates the alar and basal plates.

Liver: *Figs. 23-28, 43, 44, 72-82*; digestive and endocrine gland, which develops as an evagination of the GI tube by about 30 days; the liver becomes an important temporary blood forming organ during development until the bone marrow can take over as the adult hematopoietic tissue; furthermore, blood entering the embryo from the maternal system enters the liver first, and many metabolites are transformed before being sent to other portions of the embryo.

Lumbosacral Plexus: *Figs. 46, 86*; (plexus: braid); network of nerves supplying the leg; develops by 38 days.

Lung: *Figs. 44, 45, 67-77*; organ of gaseous exchange; does not function until birth, but appears by 30 days.

M

Mammary Ridge: *Fig. 29*; ridge of tissue between the arm and the leg buds (appearing at about 30 days) from which the mammary glands develop beginning at about 40 days; abnormal, ectopic breasts usually develop along this ridge.

Mammillary Body: *Fig. 52*; (mammillary: nipple); small bilateral projections from the posterior hypothalamus; develops from the diencephalon as a recess at about 35 days.

Mantle Layer: *Figs. 12, 16, 49, 50*; middle layer of the developing brain; at 50 days it accounts for about 1/3 of the thickness of the cortex, at 70 days it accounts for 75 percent of the total thickness, and 90 percent by 100 days; it continues to expand and accounts for almost all of the cerebral cortex.

Marginal Layer: *Figs. 12, 16, 49, 50*; outer layer of the brain; accounts for about 20 percent of the cortex at 50 days, but it decreases to only a very thin outer layer of the adult brain.

Marginal Vein: *Figs. 23, 26, 27, 46*; the primordial vein of the developing limb appearing by about 35 days and gives rise to the cephalic and basilic veins by about 45 days.

Meckel's Cartilage: *Figs. 18, 40, 59-63*; (mandibular); a temporary bar of cartilage in the developing mandible (appearing at 40 days); later replaced at about 80 days by dermal bone in the adult mandible.

Mesencephalon: *Figs. 12-16, 50*; the middle portion of the developing brain beginning at 22 days; it develops into the midbrain with tegmentum and colliculi.

Mesenteric Artery and Vein: Inferior: *Fig. 46*; **Superior:** *Figs. 28-31, 45, 81*; main vessels to the intestines; first appear as ventral intersegmental arteries from the paired dorsal aortae and supplying the yolk sac by about 22 days; as the yolk sac is absorbed into the gut, these arteries fuse into the omphalomesenteric, which becomes the superior mesenteric by 30 days; the inferior mesenteric develops directly from the ventral intersegmentals at about 30 days.

Mesonephric Duct: *Figs. 26-33, 46, 85-87*; duct of the mesonephros kidney; begins to develop in the superior region of the embryo (about 23 days) and progresses inferiorly reaching the cloaca by about

25 days; the metanephric duct develops from the inferior region of the mesonephric duct at about 28 days; in males the mesonephric duct becomes the vas deterens, in females it degenerates (by 90 days).

Mesonephros: *Figs. 10, 25, 32, 33, 46, 83, 84*; (mesonephros: middle kidney); the intermediate stage of kidney development; begins to develop at about 25 days and develops in an anterior-posterior direction, development and function reach a peak at about 45 days, and is degenerating by 50 days; by 90 days it has been completely replaced by the metanephros.

Metanephric Blastema: *Fig. 34*; (metanephric: last kidney; blastema: sprout); the beginning bud of the final (adult) kidney. (See Kidney).

Metencephalon: *Figs. 12-14*; hindbrain; first develops at about 24 days as the neural tube closes; the cerebellum begins to develop in this region by about 40 days; the pons develops from the floor of the metencephalon and the cavity (metacoele) becomes the 4th ventricle.

Mitral Valve: *Figs. 22, 71;* (mitral: turban; shape of a bishop's miter); the bicuspid or left atrioventricular valve; begins to develop at about 35 days.

Müllerian Duct: *Figs. 85, 86;* the female reproductive ducts; begin to develop in both males and females at about 45 days; in the male they soon degenerate; in the female they persist as the uterine tubes; the inferior Müllerian ducts meet at about 48 days and fuse to form the uterus at about 55 days.

Muscles: the muscles develop from the myotomes of the somites; the somites begin to develop at 19-20 days and are all present by 30 days; the muscle cells begin to migrate very shortly after somite formation; the limb muscles are present by 37 days and, after considerable reorganization, the adult configuration (even the intrinsic hand muscles) is reached by about 65-70 days.

Abdominal: *Figs. 73-79; 84;* three layers lateral (two oblique and a transverse) and two layers medial (rectus and transverse); muscles which develop as fusions of adjacent myotomes; the tendonous inscriptions of the rectus absominus are reminiscent of the earlier metamerism.

Adductor: *Figs. 86-88;* (adduction: toward); a group of muscles in the medial portion of the leg which pull the leg in toward the midline.

Biceps Brachii: *Figs. 65-67;* one of the main flexors of the elbow; located in the anterior compartment of the arm.

Biceps Femoris: *Fig. 89;* one of the hamstring group; in posterior thigh; flexes the knee.

Brachialus: *Fig. 67;* elbow flexor in the anterior arm.

Brachioradialus: *Figs. 66-67;* elbow flexor in the anterior lateral portion of the forearm.

Capitis: *Figs. 60, 63, 65;* muscles of the posterior neck which extend the neck.

Deltoid: *Figs. 64-65;* triangular muscle of the shoulder which abducts the arm.

Erector Spinae: *Figs. 74, 81-84;* muscles of the back responsible for maintaining the erect stature of the adult.

Extensor Carpi Digitorum: *Figs. 65-67, 69;* extensors of the wrist located in the posterior compartment of the forearm.

Flexor Carpi Digitorum: *Figs. 66-68;* flexors of the wrist located in the anterior compartment of the forearm.

Gastrocnemius: *Figs. 89, 90;* one of the major calf muscles in the posterior leg; flexes the foot.

Gluteal Muscles: *Figs. 86-90;* muscles of the posterior hip; extend and abduct the thigh.

Intercostal: *Figs. 68, 76, 80;* segmental muscles between the ribs.

Latissimus Dorsi: *Figs. 67-71;* large muscle of the back which extends the arm.

Masseter: *Fig. 61;* muscle of the jaw; major muscle of mastication.

Pectoralis Major, Minor: *Figs. 65, 68;* muscles of the anterior thorax. **Major:** flexes arm; **Minor:** draws scapula down.

Pronator Quadratus: (rectangle); **teres** (round), *Figs. 65, 66, 67;* muscles of the anterior forearm; pronate the hand.

Rectus: *Figs. 59-60;* (eye); (rectus: straight) four muscles around each eye; pull the eye to see up, down, or to the side.

Semimembranosus, Semitendinosus: *Figs. 88, 89;* hamstring muscles of the posterior thigh; flex the knee.

Soleus: *Fig. 90;* calf muscle of the posterior leg; flexes the foot.

Sternocleidomastoid: *Figs. 60, 62, 63;* muscle from the clavicle and sternum to the mastoid process of the temporal bone; turns the head.

Temporal: *Fig. 58;* muscle of mastication attached to the jaw.

Tibialis: *Fig. 90;* anterior: extends foot; posterior: flexes foot.

Trapezium: *Figs. 63, 68;* muscle of the back; rotates the scapula.

Triceps Brachii: *Figs. 67-70;* muscle of the posterior compartment of the arm; extends the forearm.

Vastus: *Figs. 86-88;* three of the four quadriceps group in the anterior thigh; extend the knee.

Myelencephalon: *Figs. 12-15, 38, 55-56;* (myelen: medulla); that portion of the primitive brain which will become the medulla oblongata; develops as the last six neuromeres at about 25 days.

Myotome: *Figs. 16, 18, 20, 25, 26, 36;* one of the three divisions of each somite; gives rise to the body musculature. (See Muscles.)

N

Nasal Cavity: *Fig. 61;* develops as an enlargement of the nasal pit by about 40 days; connects with the oral pharynx through the nasal pharynx.

Nasal Pit: *Figs. 20, 21, 41;* early invagination of the nasal placode; first appears by about 32 days; it ends blindly; enlarges to form the nasal cavity.

Nasal Process: *Figs. 21, 41;* tissue outgrowths from the face which begin to develop by 35 days and join together by 45 days to form the tissue of the external nose.

Nasal Septum: *Figs. 60-62;* vertical wall separating the left and right nasal cavaties; present from the time that the nasal pits are first formed at 32 days, but not delineated until the pits deepen into the nasal cavaties and move toward the midline at about 40 days; the septum then elongates ventrally meeting the palatine shelves as they fuse at about 60 days.

Neural Crest: *Figs. 6, 8;* crest of the wave of the neural folds as they move toward each other to fuse in the midline between 18 and 25 days; cells from this crest migrate away from the midline to form the dermal bones of the skull, peripheral nerves and ganglia, and pigment cells.

Neural Tube: *Figs. 7–10;* the fluid-filled tube which will give rise to the central nervous system (CNS); derived between 18 and 27 days by enlargement and fusion of the bilateral neural folds; it consists of three layers—**ependymal,** where cell proliferation occurs; **mantle,** which becomes the cerebral cortex; and **marginal**; the central neurocoele cavity becomes the brain ventricles and spinal cord canal.

Neuromeres: *Fig. 14;* early metameric divisions of the brain.

Nodosal Ganglion: *Figs. 17, 60;* the inferior ganglion of the vagus nerve (X).

Notochord: *Figs. 6–9, 18, 21, 36, 67, 75, 87;* the first structure to be laid down during development; it consists of a rod of cells running from just behind the hypophysis to the tip of the spinal column; it is laid down by the anterior tip of the primitive streake (Hensen's node) as it regresses between 15 and 25 days; influences neural tube closure, somite chondrogenesis, and perhaps body metamerism.

O

Oculomotor Nerve (III): *Figs. 15, 52–57;* nerve to the extrinsic eye muscles; appears at about 30 days.

Omphalomesenteric Vein: *Fig. 45;* continuations of the vitelline veins (which are on the yolk sac) into the body proper; they develop by 22 days, anastomose between 30 and 35 days, and give rise to the hepatic portal vein by 45 days.

Optic Cup: *Fig. 18.* (See Eye.)

Optic Nerve (II): *Figs. 18, 59;* nerve carrying sensory impulses from the eye to the brain; develops from the optic stalk. (See Eye.)

Oral Cavity: *Figs. 40, 63;* mouth; anterior end of the digestive tract; develops as a stomodeal invagination (ectoderm), which develops by 23 days; the oral plate (tissue between stomodeum and foregut) begins to break down at about 26 days opening an anterior passage from the exterior into the digestive tract; the oral cavity is continuous with the pharynx.

Otic Vesicle: *Figs. 14, 15, 39;* the early primordia of the inner ear, which develops by about 28 days; divides into vestibular and cochlear portions by about 30 days; the evagination of the endolymphatic duct also develops at about 30 days; the vestibule begins to form into semicircular canals by about 37 days and the cochlea begins to curl at the same time; by 42 days the center of each semicircular canal has been reabsorbed leaving the canals connected to each other only at the ends; the cochlea continues to curl and reaches its final form by 55 days.

P

Palatine Shelf: *Figs. 60–62;* bilateral primordia of the hard palate; they begin to develop as vertical projections from the maxilla by 42 days; they continue to elongate and then swing to the horizontal, beginning to fuse by 60 days, but fusion (including soft palate fusion) is not complete until about 90 days.

Pancreas: *Figs. 81–82;* **Dorsal:** *Figs. 28, 81;* **Ventral:** *Figs. 28, 81;* the gland that secretes digestive fluids into the duodenum and insulin and glucagon into the blood; at about 30 days the ventral pancreas develops in conjunction with the liver primordium and the larger dorsal pancreas develops on the opposite side of the developing duodenum; the ventral pancreas swings around the duodenum by 50 days and fuses with the dorsal pancreas; the dorsal pancreas becomes the bulk of the adult pancreas while the ventral pancreas contributes the proximal portion of the duct.

Pericardial Cavity: *Figs. 19-21, 23, 41-44, 66, 71-73;* the cavity containing the heart; begins to develop by about 25 days.

Peritoneal Cavity: *Figs. 23, 30-32, 73-79, 86, 88;* abdominal cavity; separated into a cavity at about 40 days.

Petrosal Ganglion (IX): *Fig. 16;* the inferior ganglion of the glossopharyngeal nerve.

Pharyngeal Pouch: **First:** *Figs. 6, 16, 17;* **Second:** *Fig. 17;* **Third:** *Fig. 18;* **Fourth:** *Fig. 19;* evaginations of the lateral walls of the pharynx; they develop by 30 days and are gone by 40 days; the 1st pouch becomes the auditory tube, the 3rd becomes the thymus.

Pharynx: *Figs. 6, 17-19, 40, 59-62;* the throat; develops as the foregut by 20 days; becomes connected with the outside through the mouth by about 26 days and becomes connected with the nasal cavity by 40 days.

Phrenic Nerve: *Figs. 23, 66, 70;* nerve from the cervical plexus to the diaphragm; begins to develop by about 35 days.

Pia Mater: *Figs. 12, 51;* the inner most layer of meninges surrounding the central nervous system.

Pigmented Retinae: *Figs. 6, 18, 40,59; 60;* the outer most retinal layer of the eye; inhibits light reflection; develops from the outer layer of the optic cup.

Pleural Cavity: *Figs. 23, 73-75;* cavity of the lungs; separated as a cavity by 40 days by formation of the pleuroperitoneal membrane.

Pons: *Figs. 38, 52-56;* (Pons: bridge); swelling at the inferior, ventral portion of the brain stem; develops from the floor of the metencephalon.

Posterior Communicating Artery: *Figs. 15, 17, 38, 53-56;* bilateral arteries developing from the internal carotid arteries at about 27 days along the base of the diencephalon; they fuse with the basilar artery at about 32 days to form a portion of the circle of Willis.

Pulmonary Artery: *Figs. 41, 69, 70;* arteries carrying blood to the lungs; develop from the 6th aortic arch by 30 days.

Pulmonary Trunk: *Figs. 20, 68, 69;* artery leading from the right ventricle to the pulmonary arteries; separated from the aorta by formation of the spiral septum at about 35 days.

R

Rathke's Pocket: *Figs. 6, 17, 39;* evagination of the foregut which becomes the adenohypophysis (see Hypophysis); develops at about 30 days.

Rectum: *Figs. 32, 85-89;* the most posterior portion of the digestive tract; develops as part of the hindgut at 30 days; is formed by the separation of the cloaca into rectum and urogenital sinus at 34 days.

S

Sclerotome: *Figs. 16, 18, 20, 25, 35;* one of the three divisions of each somite; gives rise to the vertebrae and ribs.

Semilunar Ganglion (V): *Figs. 15, 16, 39, 58;* ganglion of the trigeminal nerve.

Sensory Retinae: *Figs. 6, 18, 40, 59, 60;* inner layer of the optic cup which gives rise to the rod and cone layer and nervous layers of the adult retinae.

Septum Primum: *Figs. 21, 43, 71;* first interatrial septum (left of the two septa); begins to develop at about 28 days and closes the foramen primum at about 40 days as the foramen secundum forms by resorption of part of the septum primum; functions as a movable flap over the foramen ovalle until birth.

Septum Secundum: *Figs. 21, 43, 71, 72;* second interatrial septum (right of the two septa) which begins to develop by about 45 days and continues to close leaving the foramen ovalle which remains open until the septum primum is forced against it at birth.

Sinoatrial Valve: *Figs. 8, 21, 22, 70, 71;* (valvulae venosa) valve between the sinus venosus and right atrium which forms a temporary valve in the early heart; develops by about 25 days and degenerates as the permanent valves develop by about 55 days.

Sinus Venosus: *Figs. 8, 22, 23, 70, 71;* the most posterior of the four primitive heart chambers appearing at about 22 days; it is formed by fusion of the two cardinal veins; it empties blood into the atrium; at about 55 days it is absorbed into the right atrium.

Somite: *Fig. 10;* early metamers of the embryonic axis; there are forty in number which begin to develop at 19 days and are all present by 30 days; they form three tissues—dermatome, mytome, and sclerotome, which give rise to the dermis (there is some question about this point now), muscles, and vertebrae, respectively.

Spinal Accessory Nerve (XI): *Figs. 13-15;* nerve to the muscles of the neck and upper thorax; begins to develop at about 30 days.

Spinal Cord: *Figs. 17-19, 21, 25-30, 32-36, 39-41, 43, 44, 47, 59-61, 65, 71-82, 85-88, 90, 91;* part of the central nervous system; begins to form at about 23 days and is a closed tube by 27 days. (See Neural Tube.) When it first develops it reaches the tip of the spinal column in the tail but later regresses until it only reaches to about the 3rd lumbar vertebra at birth.

Spinal Ganglion: *Figs. 18-20, 23-31, 34, 40, 45-47, 61, 64, 65, 68, 73-80, 83, 85, 89;* dorsal root ganglia located in the dorsal root of each spinal nerve; formed by migration of neural crest cells by 35 days.

Spinal Nerve: *Figs. 35, 39, 42, 43, 47, 86;* nerves supplying the body and extremities; develop from neural crest cells by 35 days; there are 32 pairs, each having a dorsal (sensory) and ventral (motor) root, which join and then give off three rami (branches), a dorsal to the back, a ventral to the body wall and extremities, and a communicating to the sympathetic nervous system.

Spleen: *Fig. 81;* lymphatic tissue which begins to develop in the greater omentum by about 40 days.

Stomach: *Figs. 26, 80-82;* first appears as an enlargement of the midgut by about 30 days.

Stomodeum: *Fig. 19;* ectodermal primordia of the mouth (see Oral Cavity); separated from the pharynx by a stomodaeal (oral) plate.

Subcardinal Vein: *Fig. 28;* veins which form along the nephric ridge by about 25 days; mostly degenerate by about 45 days; the posterior portions give rise to the suprarenal veins.

Subclavian Artery: *Figs. 22, 65, 66;* artery to the arms; the right develops from the 4th aortic arch and the left from an arterial branch of the great arch of the aorta; each develops by about 33 days.

Submandibular Gland: *Figs. 61, 62;* glands which secrete digestive juices into the mouth; develop at about 55 days.

Superior Vena Cava: *Figs. 65-68;* vein returning blood from the superior portions of the body; develops from the right superior cardinal vein by about 55 days.

Suprarenal Gland: *Figs. 78-81;* (adrenals); a gland of dual function and origin; the cortex develops from mesenchyme in the renal area beginning at about 35 days; the medulla develops from migrating neural crest cells which reach the renal area by about 45 days and invade the cortical tissue.

Sympathetic Chain Ganglion: *Figs. 20-22, 60-62, 75, 78;* paired ganglia along the sympathetic trunk connected to the central nervous system by the communicating rami and interconnecting the sympathetic nervous system.

Sympathetic Trunk: *Figs. 41, 42, 44, 63-65, 68, 73, 74, 79, 81, 84;* nerve fibers connecting the chain ganglia and thus the sympathetic nervous system.

T

Tectum: *Fig. 50;* (tectum: roof); roof of the mesencephalon from which the colliculi develop.

Tegmentum: *Figs. 14, 16;* the main portion of the mesencephalon.

Telencephalon: *Figs. 7, 21, 22;* the anteriormost segments of the brain which form with closure of the neural tube at 25 days; two lateral ventricles form at 30 days and give rise to the two cerebral hemispheres.

Thalamus: *Figs. 19, 38, 52-55;* (thalamus: bed); the major relay center of the brain; develops from the lateral wall of the diencephalon by about 40 days.

Third Ventricle: *Figs. 17, 19, 20, 38, 39, 51-55, 57;* diocoele; cavity of the diencephalon.

Thymus: *Fig. 65;* lymphoid organ in superior thorax; important in development of the immune system; develops from the 3rd pharyngeal pouch at about 40 days.

Thyroid Cartilage: *Figs. 61, 62, 69, 73, 74;* main cartilage of the larynx.

Thyroid Gland: *Fig. 63;* develops at about 27 days as an evagination from the floor of the pharynx; by 40 days the thyroid has lost its connection to the pharynx and migrates down the neck leaving the foramen caecum at the base of the tongue to mark its point of origin.

Tongue: *Figs. 40, 60-63;* develops from medial swellings of the mandibular arches as lateral lingual swellings by about 30 days; at the same time a third swelling, the tuberculum impar, forms in the midline; the three swellings fuse to form the adult tongue, with the lateral lingual swellings making up the majority.

Trachea: *Figs. 8, 20, 21, 41, 42, 64-68;* tube leading from the larynx to the bronchi; develops at about 25 days as an evagination from the upper digestive tract and quickly gives off two buds which form the lungs.

Trigeminal Nerve (V): *Figs. 39, 56, 57;* (trigeminal: threefold); sensory nerve to the face and motor nerve of mastication; develops at about 30 days.

Trochlear Nerve (IV): *Figs. 12, 13, 57;* (trochlear: pulley); nerve to the superior oblique muscle of the eye; develops by about 30 days.

Trophoblast: *Fig. 2;* (trophoblast: germ); outer layer of the blastocyst, which invades the endometrium and provides the tissue of the embryonic placenta.

U

Umbilical Artery: *Figs. 27, 30-34, 45, 46, 82-85;* paired arteries to the placenta; develops by 23 days from the most posterior ventral intersegmental arteries and are connected to the internal iliac artery later in development; degenerate after birth.

Umbilical Cord: *Figs. 27, 82;* composed of two arteries, one vein, an allantoic stalk, yolk stalk, and connective tissue (Worton's jelly); forms first as two separate stalks, the yolk and the allantoic by 20 days; they combine to form one common cord by about 30 days.

Umbilical Vein: *Figs. 27-31, 33, 34, 45, 79, 80, 82;* single vein returning blood from the placenta to the liver and thence to the heart; develops as bilateral veins associated with the allantois by 20 days; the right begins to degenerate by 33 days and is gone by 44 days; the left persists until birth when it degenerates to become the round ligament of the liver.

Urachus: *Fig. 83.* (See Allantois.)

Ureter: *Figs. 33, 34, 82, 85;* (metanephric ducts) tubes from the kidneys to the bladder; develop as evaginations from the mesonephric duct by 30 days but by 36 days they have made independent connection to the bladder.

Urethra: *Figs. 85-88;* tube connecting the bladder to the exterior; develops as part of the cloaca by 30 days; is separated off into the urogenital sinus by 36 days; and becomes the definitive urethra by 90 days.

Urogenital Sinus: *Figs. 32, 89, 90.* (See Urethra.)

V

Vagus Nerve (X): *Figs. 13, 15, 16, 18-22, 39-42, 58-65, 75, 77, 78;* major parasympathetic nerve to the neck, thorax, and upper abdominal viscera.

Ventral Aorta: *Fig. 7;* (truncus arteriosus); artery leading from the primitive heart to the aortic arches; developing by 22 days; becomes the ascending aorta.

Ventral Horn: *Figs. 21, 24, 27, 30;* accumulations of motor neurons in the ventral portion (basal plate) of the spinal cord; connected to the ventral root.

Ventral Mesogaster: *Figs. 26, 27;* (lesser omentum) the ventral mesentery of the stomach; develops into the lesser omentum by 45 days.

Ventral Ramus: *Figs. 19, 21, 22, 64;* major division of the spinal nerves carrying impulses to and from most of the body and the extremities.

Ventral Root: *Fig. 22;* division of the spinal nerve which carries efferent, motor impulses from the spinal cord; develops by 35 days.

Ventricle: *Figs. 8, 21, 23, 24, 41-43, 68-72;* one of the first four divisions of the primitive heart appearing about 22 days; divided into two chambers by an interventricular septum which begins to develop by 30 days and is closed by 40 days.

Vertebral Arch: *Figs. 59, 61, 64, 66, 67, 75;* the two major portions of each vertebra; develop from the sclerotome of two adjacent somites; the body develops around sections of the notochord beginning by 35 days; the arches begin to form by 42 days, but are not complete until about 100 days; failure of formation of the arches results in spina bifida.

Vertebral Artery: *Figs. 7, 17, 21, 39-41, 58-64;* two of the major arteries bringing blood from the subclavian arteries to the brain; develop as arteries lateral to the dorsal aortae by 25 days.

Y

Yolk Sac: *Fig. 3;* first cavity of the developing embryo; first appears as the blastocoele by 4 days; gives rise to the digestive system; the foregut and hindgut develop by 18 days.